A PHOTO COLLECTION OF REAL STARSHIPS

TERRY RAY

ROSWELL PRESS

an imprint of Sunbury Press, Inc.
Mechanicsburg, PA USA

ROSWELL PRESS

an imprint of Sunbury Press, Inc.
Mechanicsburg, PA USA

For information about special discounts for bulk purchases, please contact Sunbury Press Orders Dept. at (855) 338-8359 or orders@sunburypress.com.

To request one of our authors for speaking engagements or book signings, please contact Sunbury Press Publicity Dept. at publicity@sunburypress.com.

FIRST ROSWELL PRESS EDITION: June 2021

Set in Adobe Garamond | Interior design by Crystal Devine | Cover by Lawrence Knorr | Edited by Lawrence Knorr.

Publisher's Cataloging-in-Publication Data
Names: Ray, Terry, author.
Title: A photo collection of real starships / Terry Ray.
Description: First trade paperback edition. | Mechanicsburg, PA : Roswell Press, 2021.
Summary: Terry Ray, a veteran of MUFON, presents the many photos he has worked with to isolate what he believes are images of starships.
Identifiers: ISBN : 978-1-62006-867-0 (softcover).
Subjects: BODY, MIND & SPIRIT / UFOs & Extraterrestrials.

Product of the United States of America
0 1 1 2 3 5 8 13 21 34 55

Continue the Enlightenment!

CONTENTS

INTRODUCTION

My last UFO book, *The Complete Story of the Worldwide Invasion of the Orange Orbs,* broke new ground in the field of ufology by replacing the "conventional knowledge" at the time that orange orbs were Chinese Lanterns. The book laid out a detailed description of what these orbs are and many other aspects of the phenomenon, such as their flight patterns, their home bases on Earth, and the possible reasons they are in our skies.

The orange orb book explained that these orbs, being seen all over the world, are colorful, spherical cloaks that conceal a starship within. I included cases where observers had witnessed the cloaking process, in which a gaseous substance emerges from beneath the ship and envelops it in a spherical cloak. It also reveals that those in control of the craft can change the color of the cloaking sphere or turn off the color altogether, making it invisible in the night sky.

After writing my book, I began to wonder if it were possible to somehow see through these cloaks and observe the craft inside, so I began experimenting with photographic techniques to see if I could find a way to do it. It took almost a year before I was able to see the blurry outlines of what appeared to be some sort of a structure within the cloak. Then, bit by bit, I was able to bring more clarity to the image until I finally had a clear image of the starship in remarkable detail.

But I was puzzled by what I saw. The crafts were extremely complex but nothing like I had imagined, since I was anticipated seeing starships like those in the *STAR WARS* movies. After contemplating this for a while I realized how silly this was.

We are a civilization that has not yet learned to manipulate gravity and still rely on air pressure differentials to keep our aircraft flying and the first serious attempt to understand gravity waves will not take place until the LISA Project, set to launch in the early 2030s. Because of this, we still use airfoil wings and streamlined bodies, using wind manipulation for directional control. Not knowing what real starships look like, the creators of STAR WARS, obviously chose to use Earth-like aircraft design to design its models.

When a civilization learns to control gravity, size, shape, and weight no longer matter. With gravity control, we could build an Eiffel Tower while it floats in the air and, being weightless, it could be moved with a push of a hand. and its speed would be almost unlimited.

Also, with the proper gravity control within the interior of the craft, the occupants would not be affected by the inertia of acceleration, deceleration, or changes in direction. The craft could stop on a dime and dramatically change direction without the occupants feeling any of it.

To picture this, think of yourself sitting in a movie theater with a 360-degree screen surrounding you. The screen could create images of moving forward at a tremendous speed, stopping instantly, and making 90-degree turns. You would see all this happening, but you would have no physical sensation of any movement. If your eyes were closed, you would not know you were moving or changing directions. This is how it would be as a passenger in a starship.

As for the photos in this collection, each one began with a photo of a hazy light in the sky—usually at night because the glowing orbs are much easier to see against a black sky. The clarity process then begins. Each one is different. Some orbs reveal their craft after fifteen minutes of clarification—some can take several hours to complete.

Following are some examples of various stages of the process. To reach each stage takes a lot of time, trial and error, and some degree of artistic judgement.

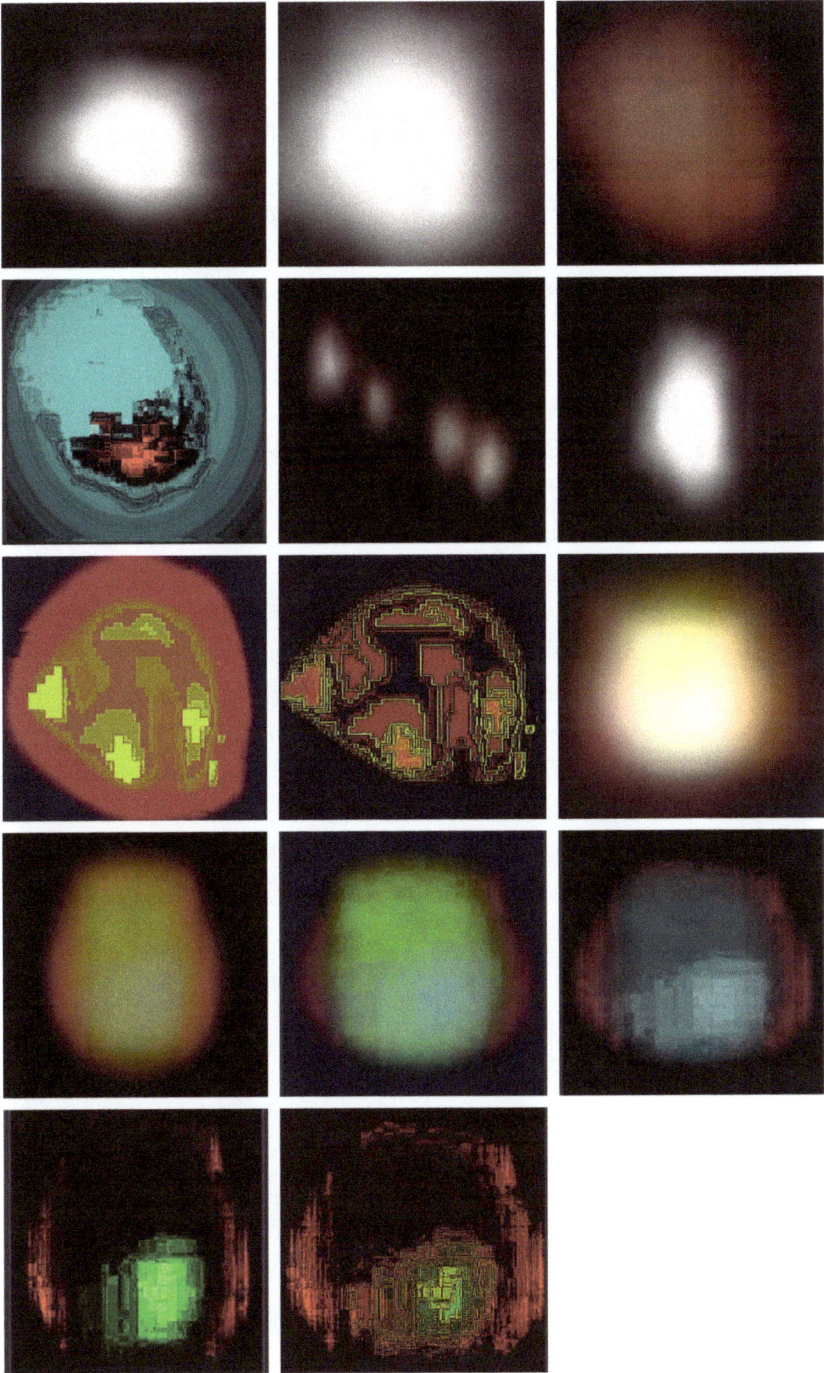

As is obvious from the clarified starships above, they are quite different from one another in both color and design. I have never clarified any ships that were identical. Some are similar but none were exactly the same. Even when the ships are in formation, each one is different, as will be demonstrated in one of the following photos. This is very un-Earth like. When our military planes fly in formation, they are normally all identical aircraft. There are, however, general categories of starships which have similar designs and this photo collection will be divided into those broad categories.

The fact that there are general categories of starships begs the question as to what this means. If we apply our Earth experiences to this question, we could come to a possible conclusion that each design category could be from a specific interstellar civilization, much like American fighter jets are designed differently than those made in Russia, China, France, or the U.K.

As a final word . . .

I anticipate this photo collection may garner some detractors who will say that these photos are nothing more than "a bunch of random pixels," but any objective observer can note the detail and intricate construction design in all the photos and see they are not at all random. I had detractors with my book on orange orbs as well but what is set forth in that book has become the new conventional view on what these orbs are. Hopefully, this book will change the current version of starships (as provided by Hollywood) to a realistic view.

THE
LUBBOCK LIGHTS

Although the "Phoenis Lights" are more famous than the lights photographed over Lubbock, Texas, on the night of August 31st 1951, the Lubbock lights are far more dramatic. Four professors from the Texas College of Technology witnessed V-shaped formations of lights more than ten times beginning on August 25th and for the next two weeks.

They reported that the formations moved across the night sky from north to south, traversing the entire sky within three seconds and in complete silence. Some of the formations had as many as 30 orbs of light. Frank Gore Jr. was lucky enough to catch the formation on camera on August 31st.

This particular formation contained 18 lights and is considered by some to be the largest V-shaped formation ever photographed. It is much larger than the Phoenix Lights, which most witnesses described as having four or five lights. What distinguished the Phoenix Lights was the number of witnesses seeing it, numbering in the multiple thousands. Most ufologists believe that more people saw the Phoenix Lights that any other UFO sighting in modern history.

Following the photo of the Lubbock Lights is a partially clarified photo of the lights. I you look closely, you will see that each of the starships is different—similar but different The next set of photos shows fully clarified pictures of each of the 18 ships in the formation.

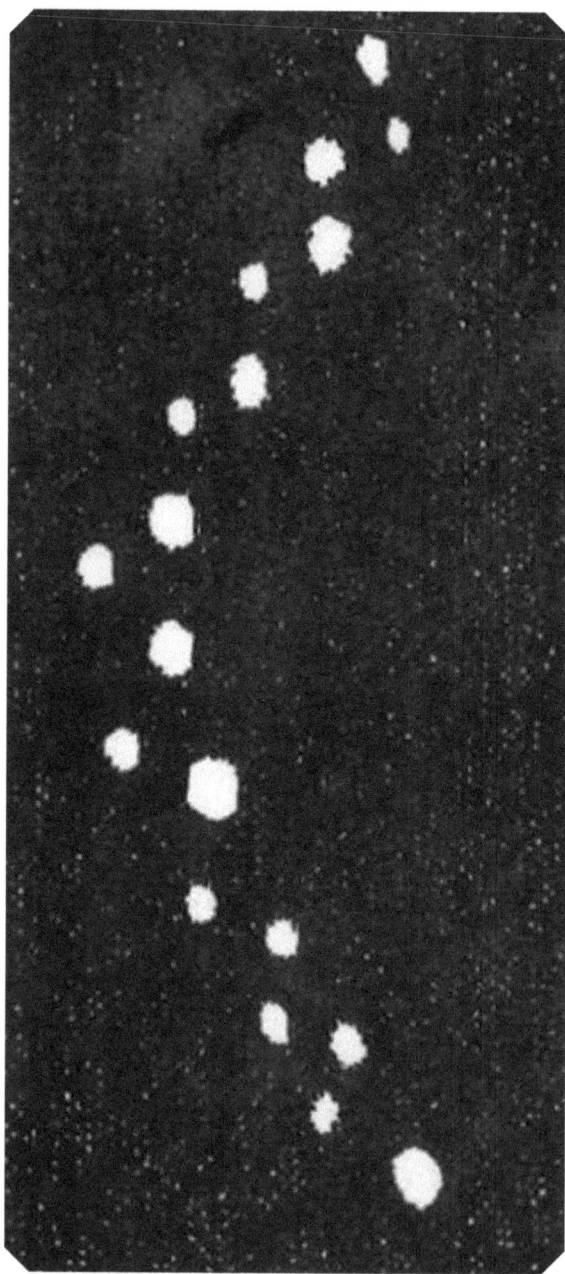

The Lubbock Lights

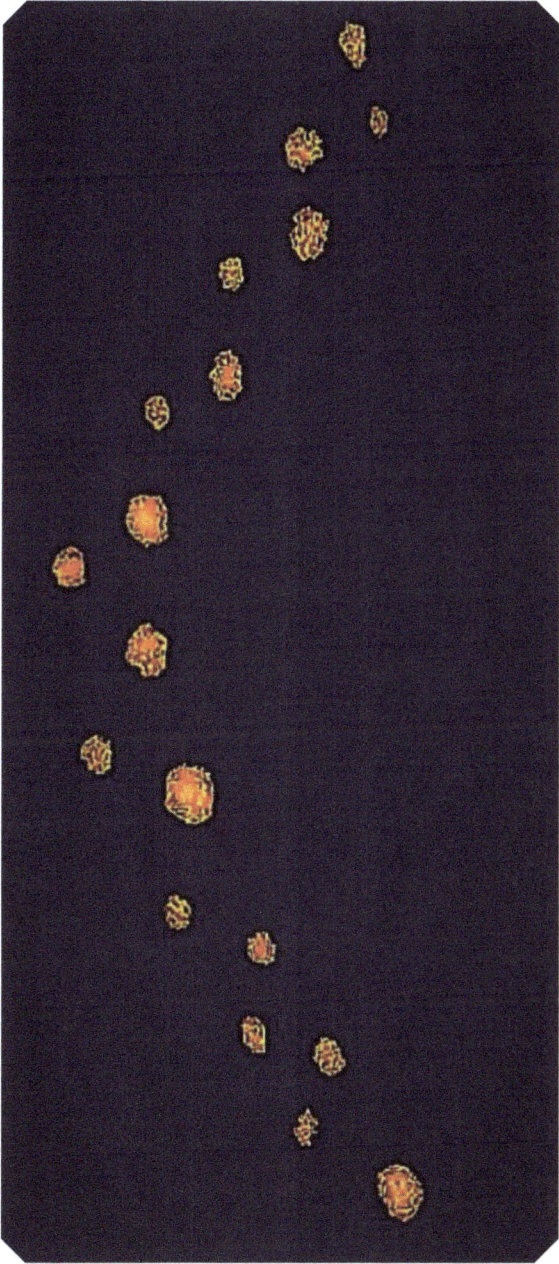

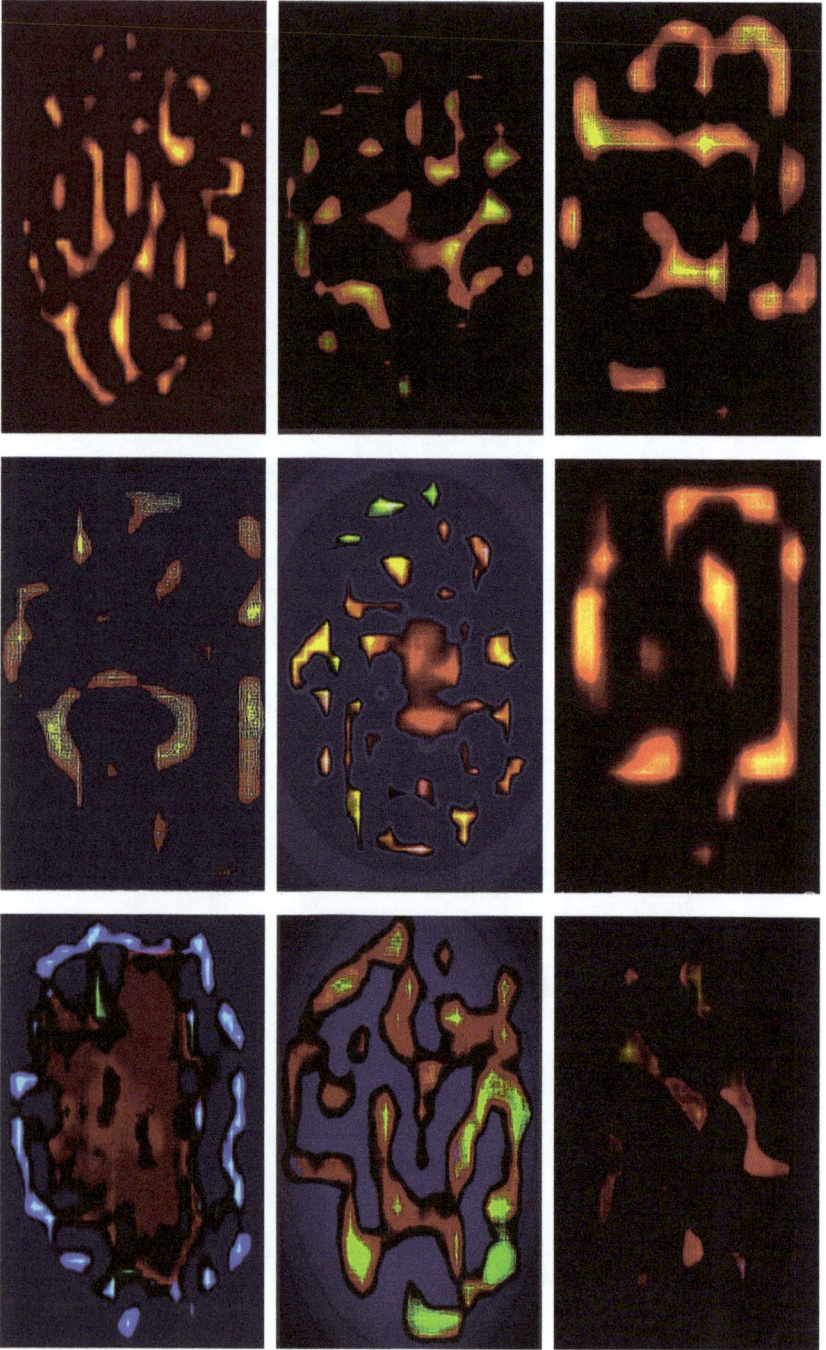

As the reader can see, most of the ships are comprised of smaller

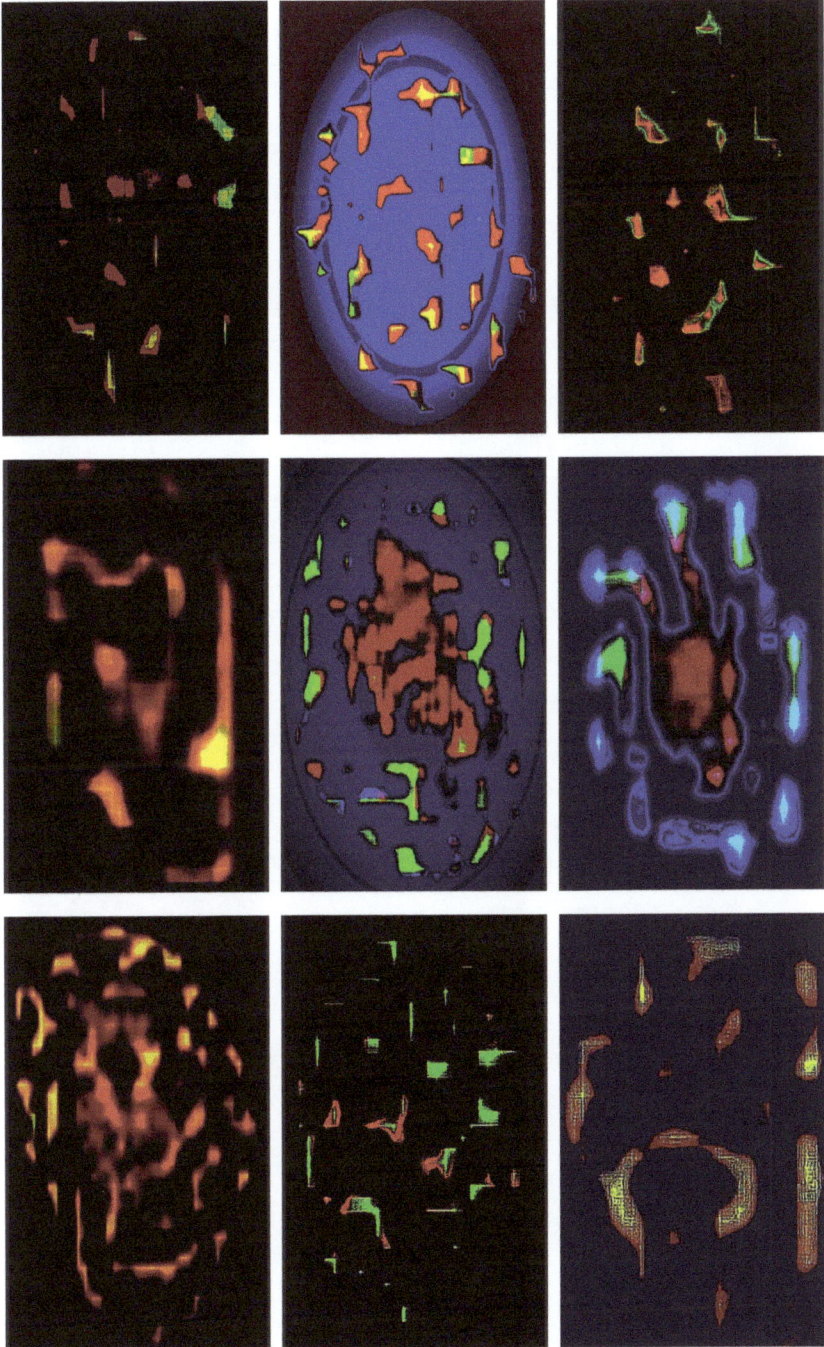

units, but four of them are quite different from the rest.

CATEGORY ONE

ROUND AND OVAL STARSHIPS

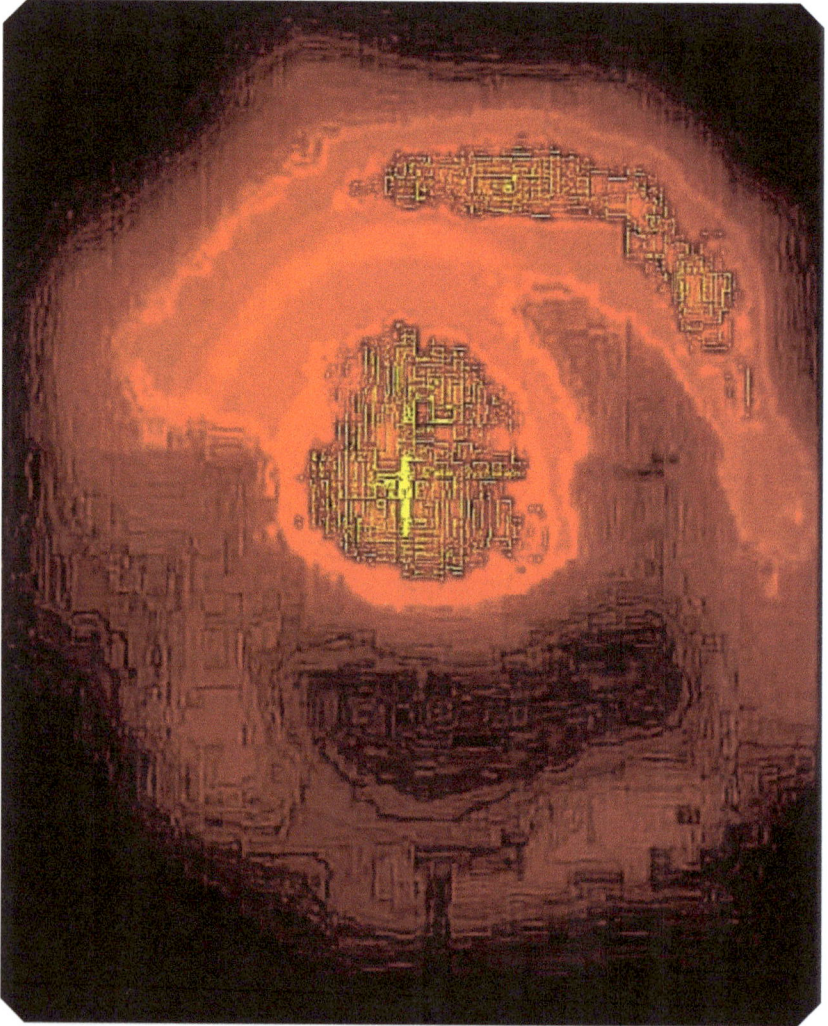

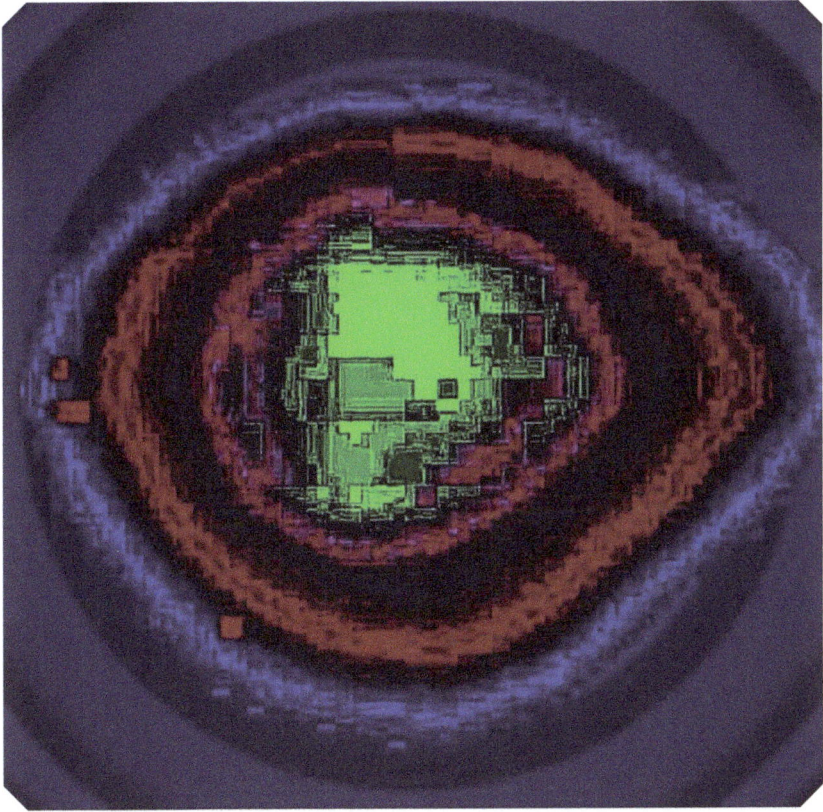

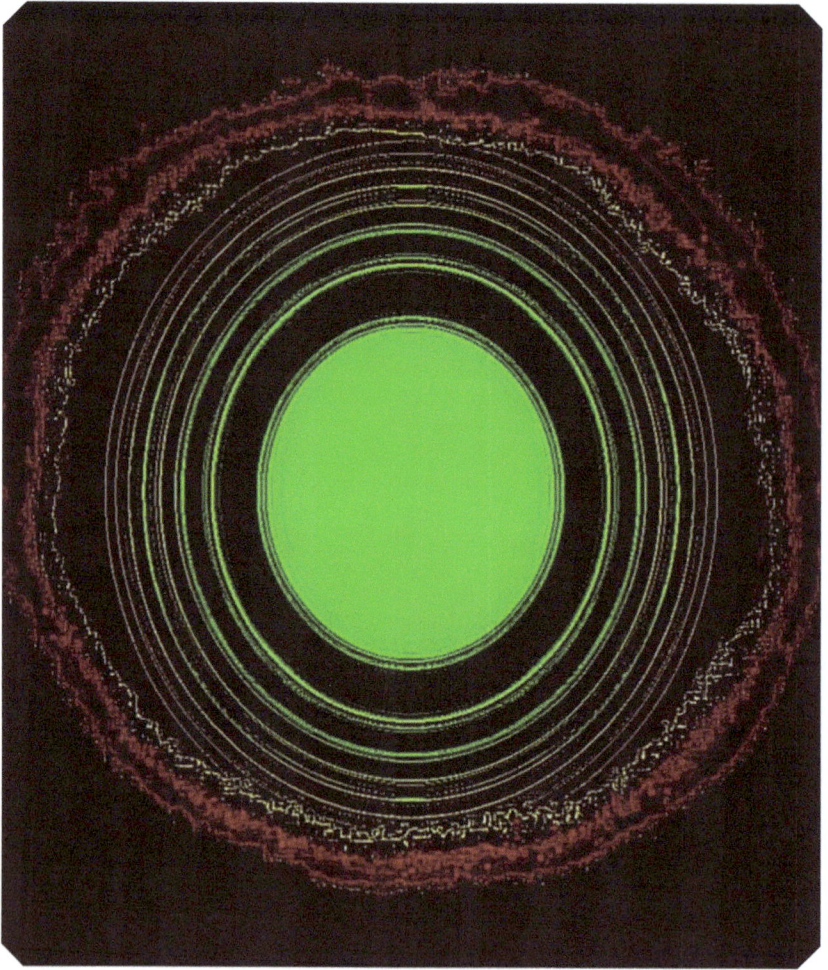

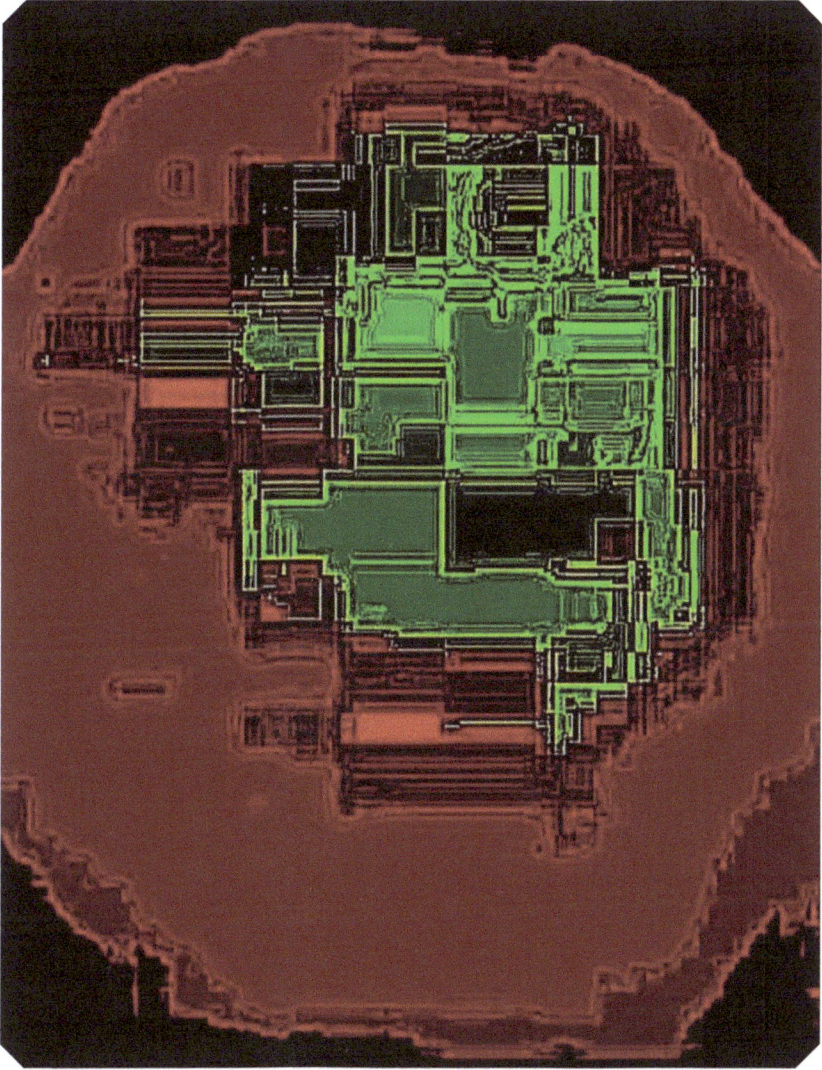

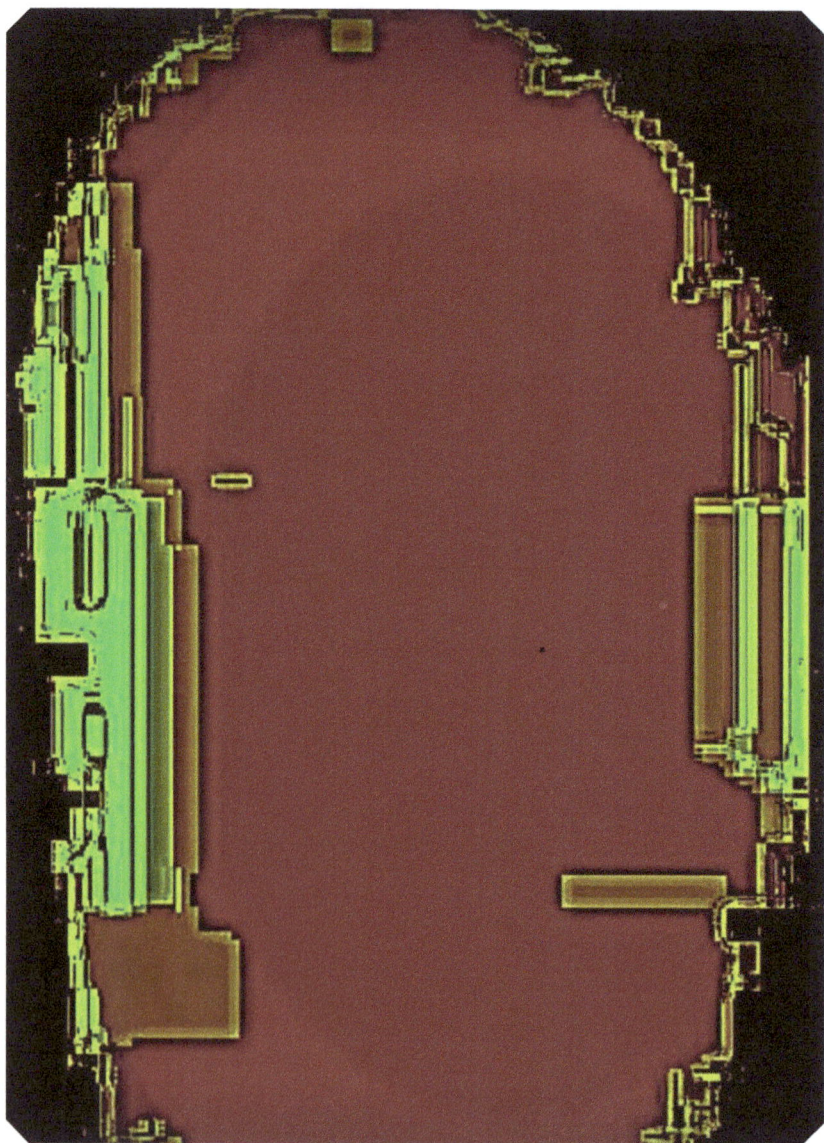

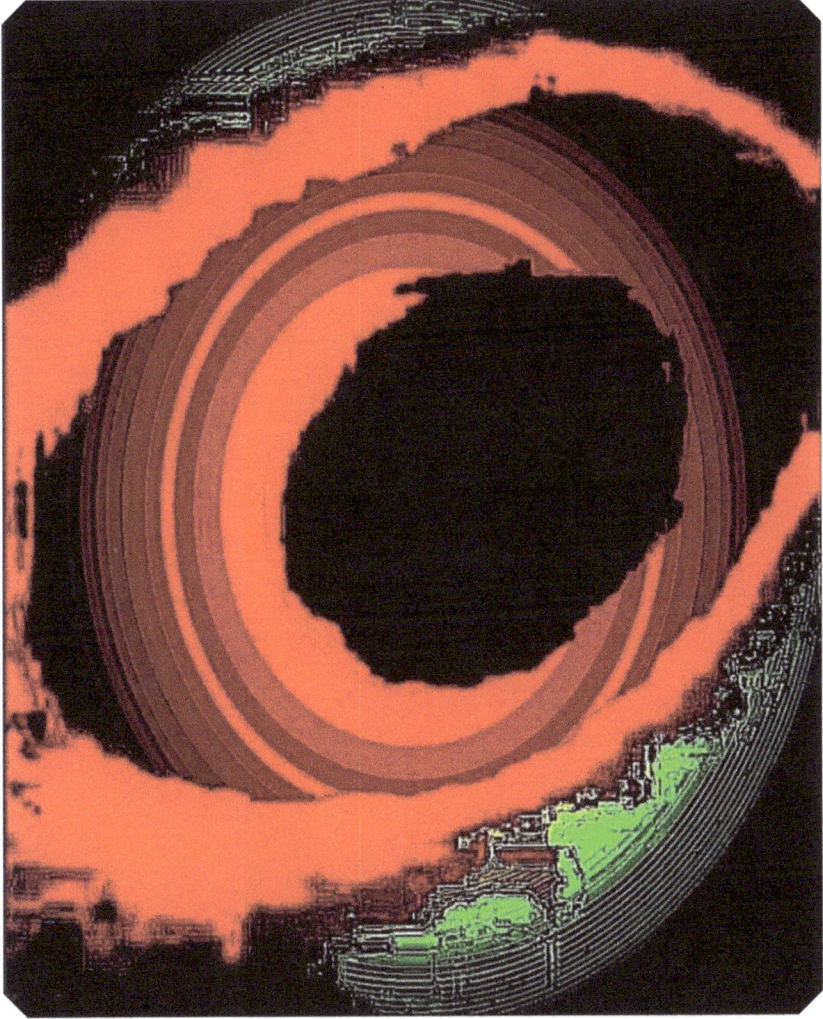

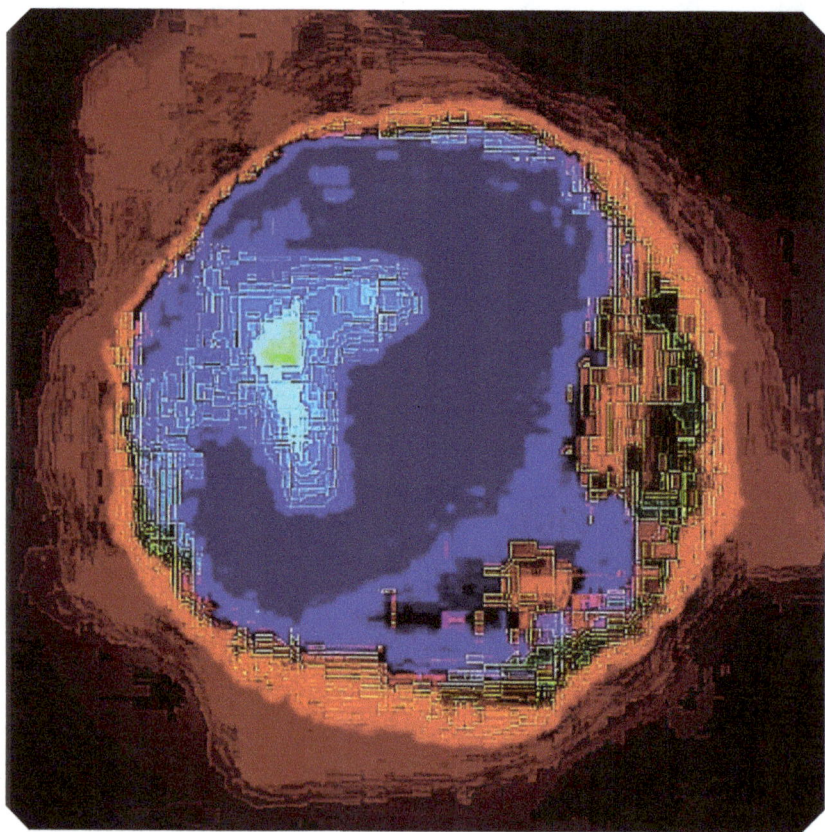

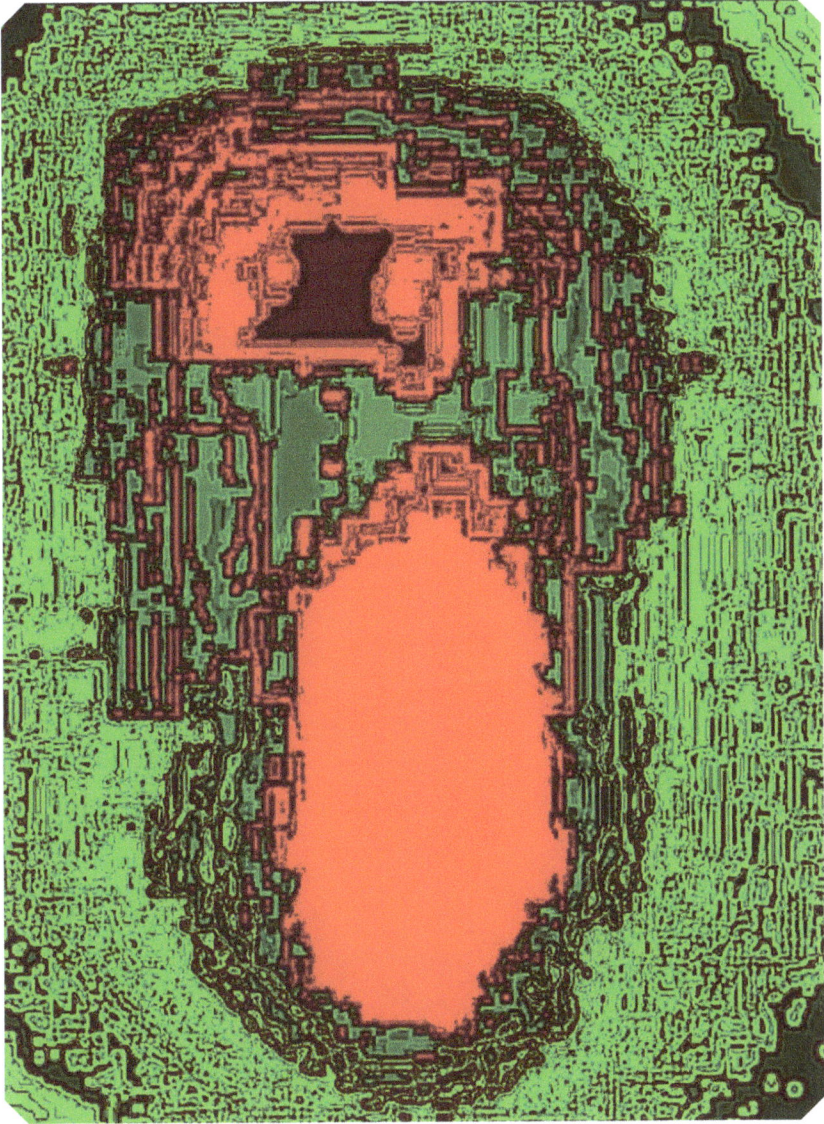

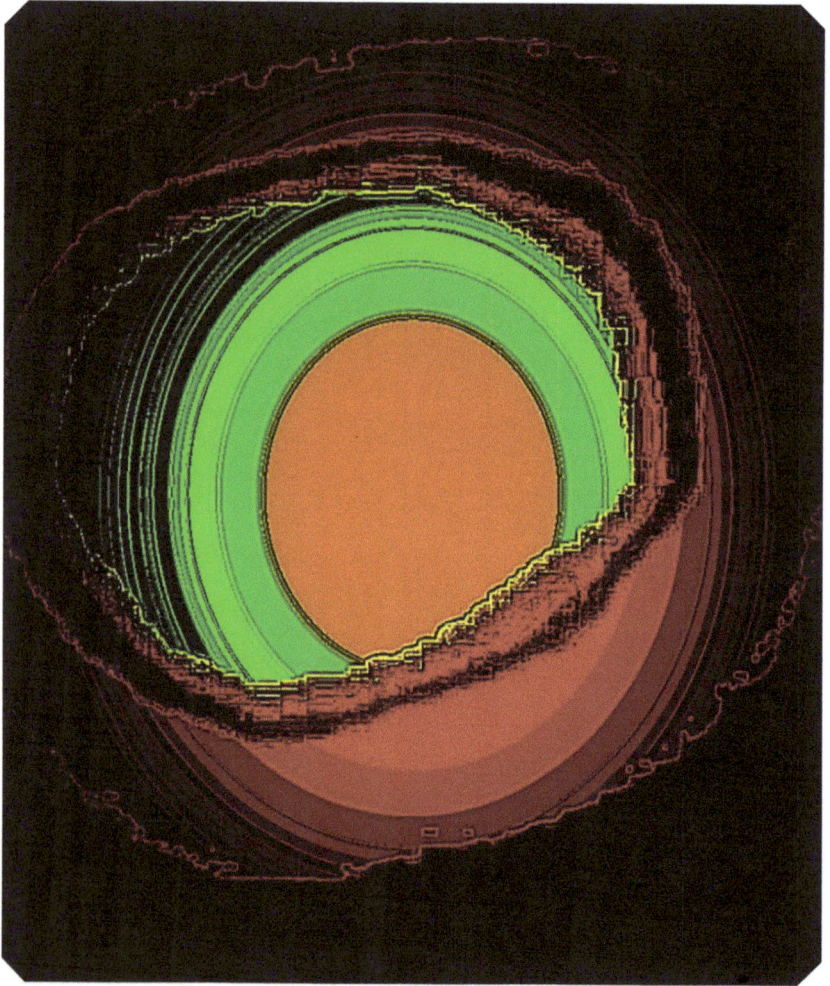

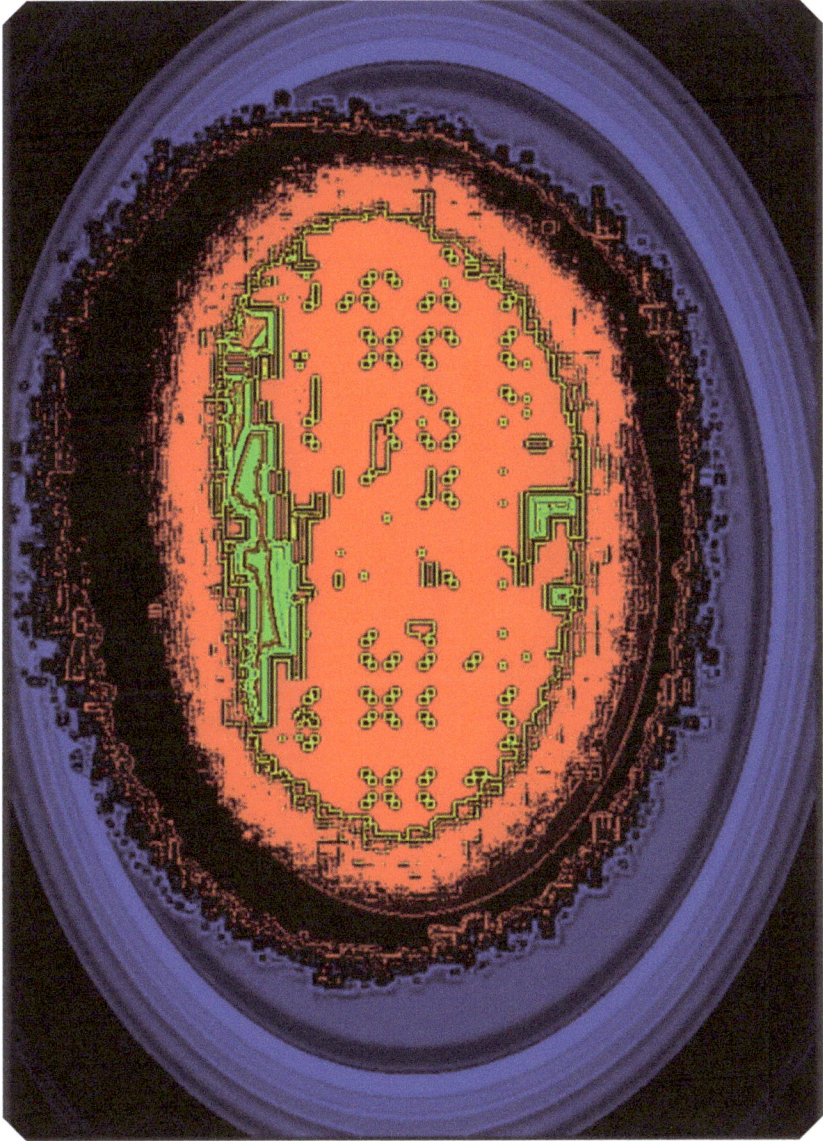

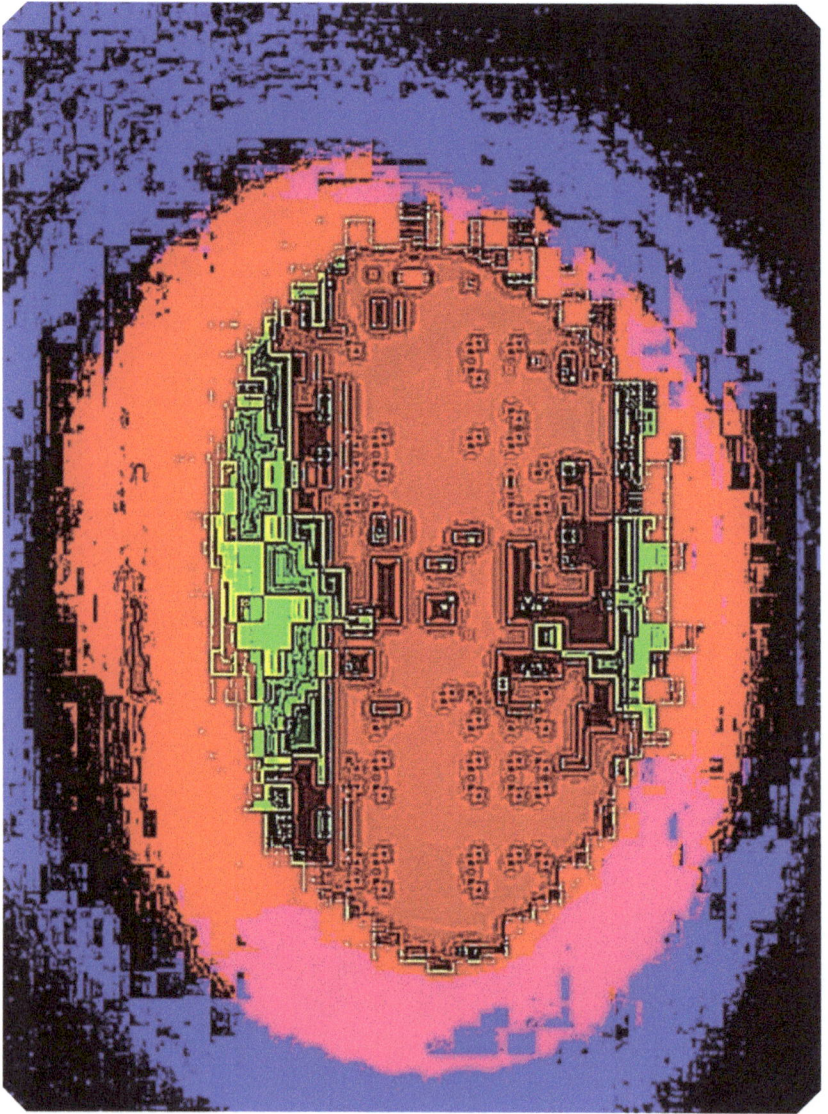

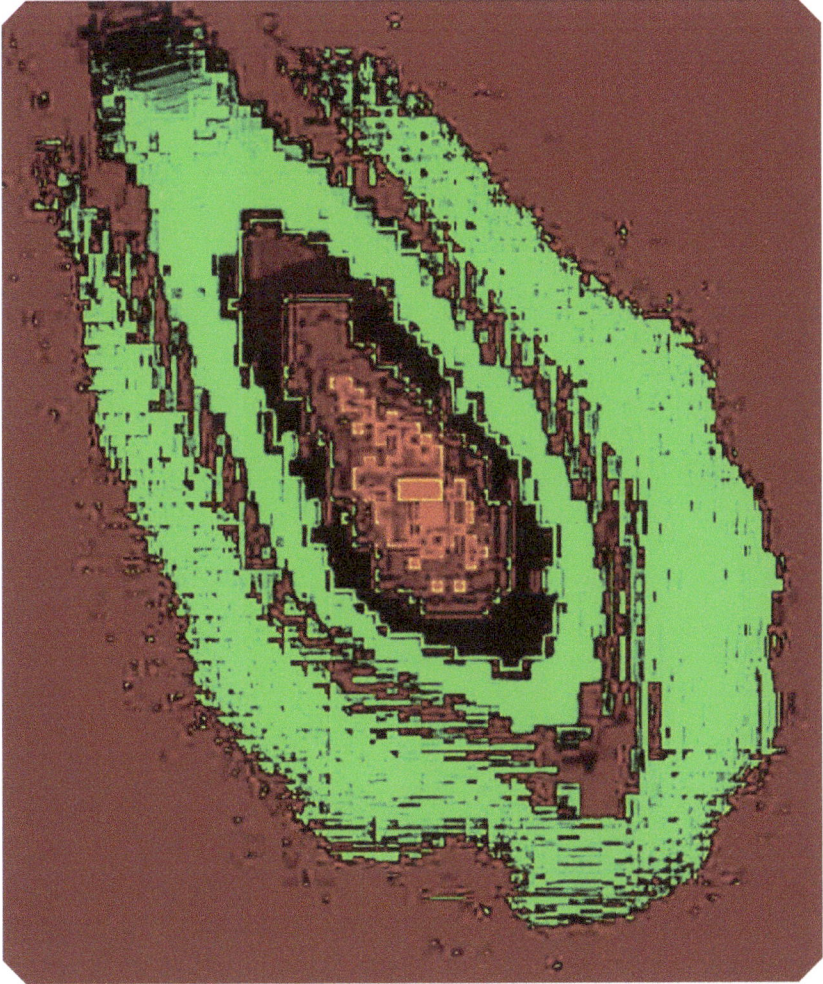

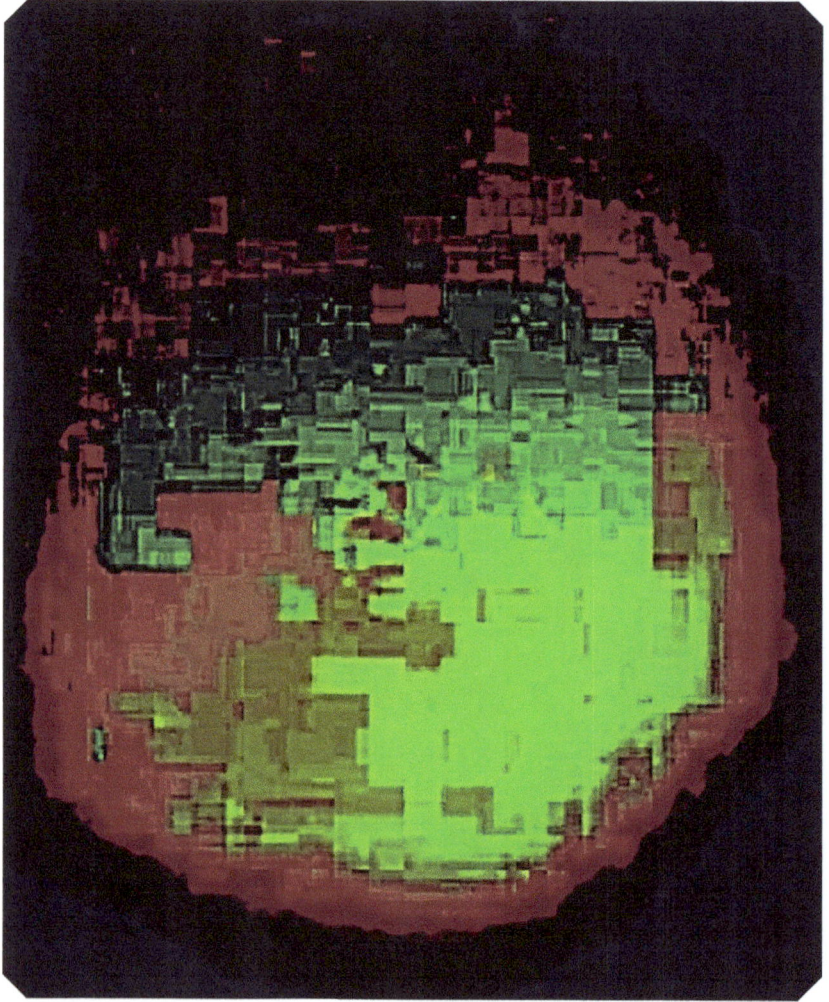

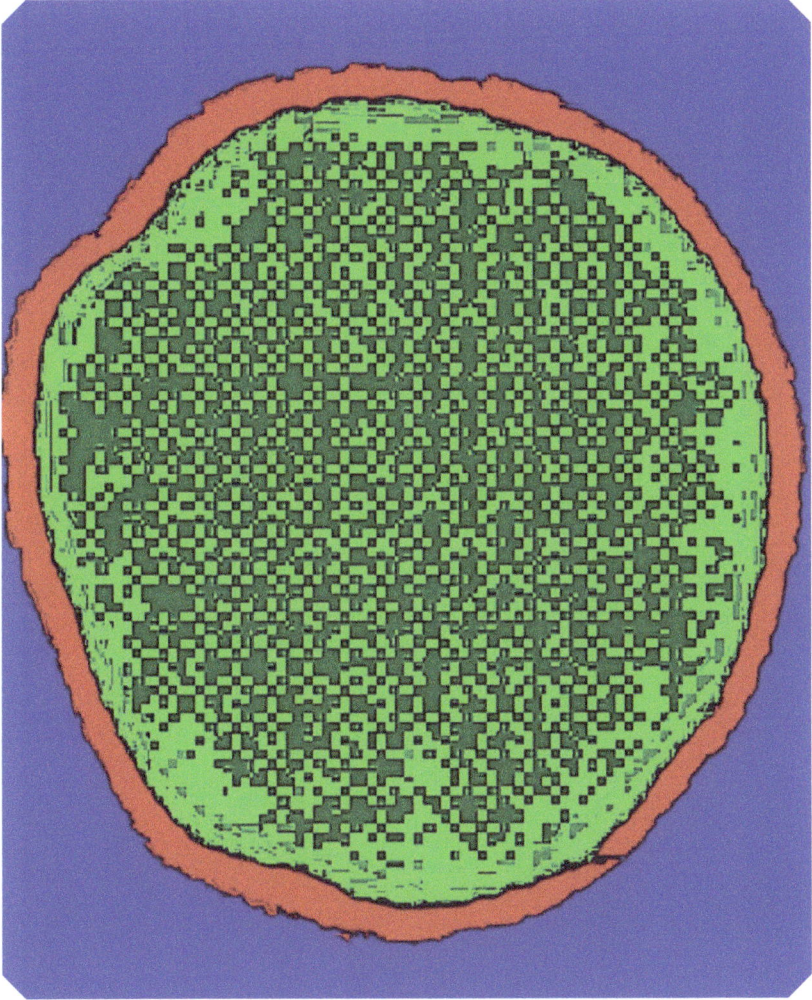

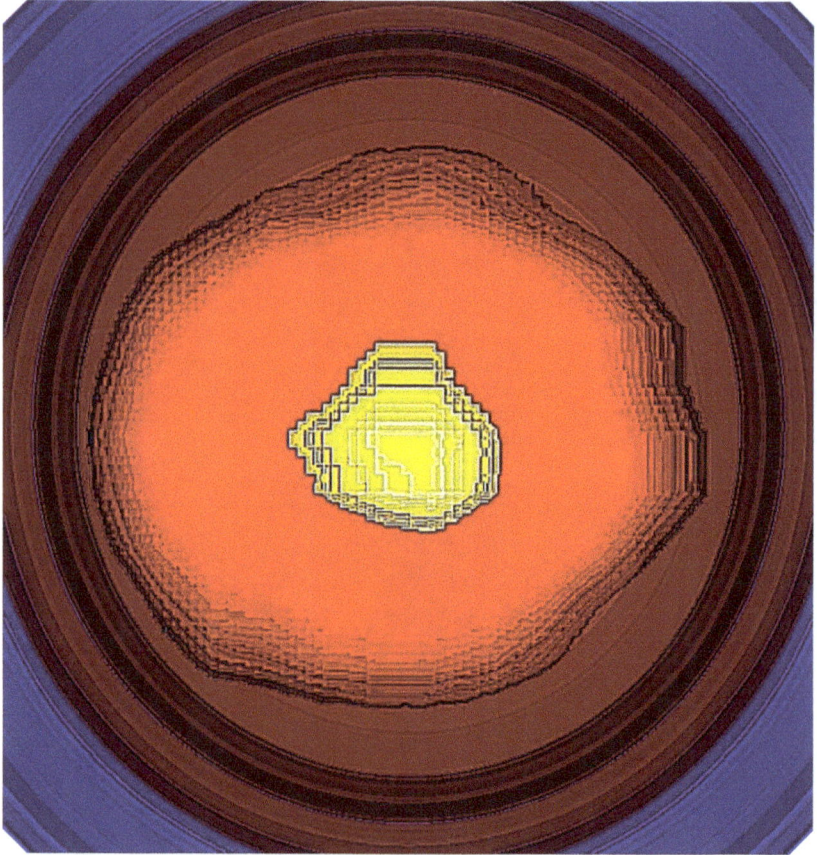

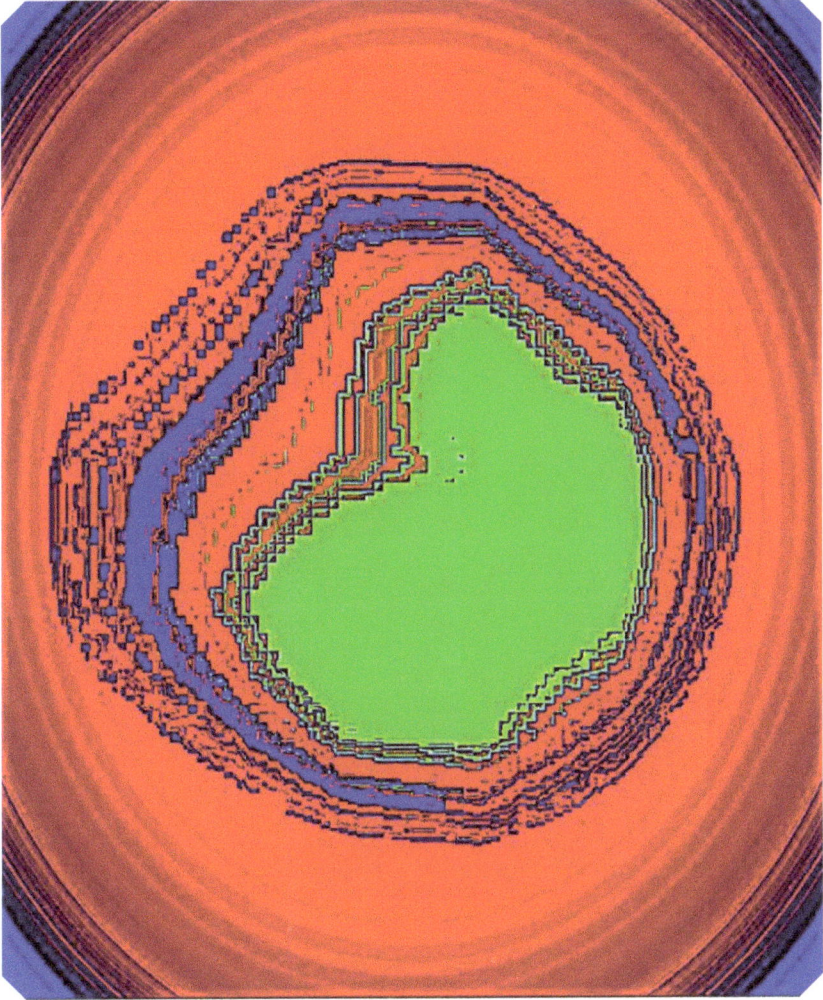

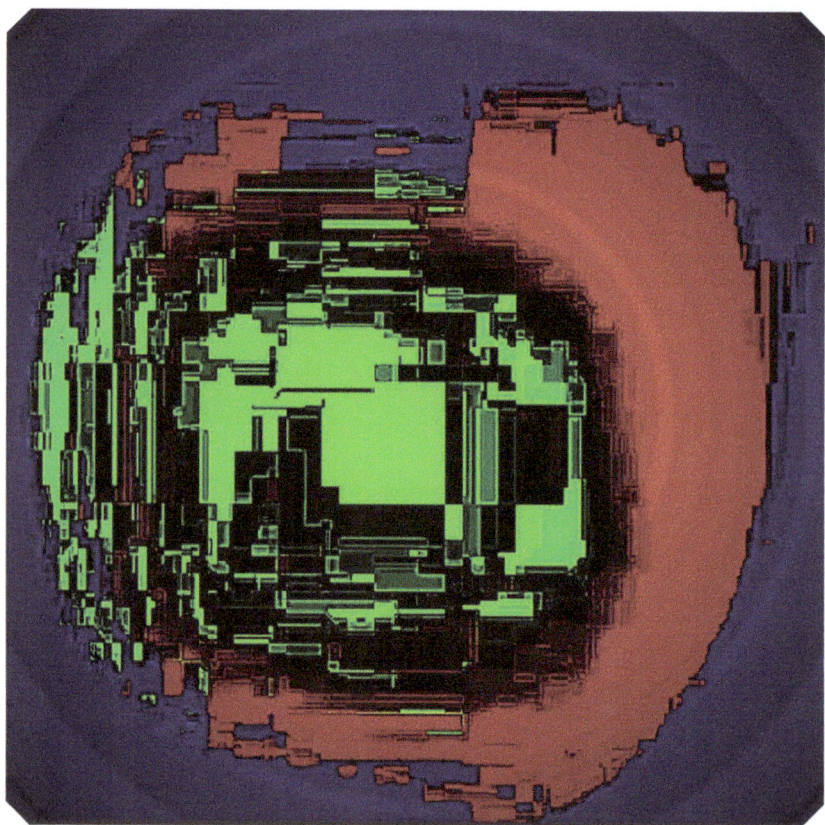

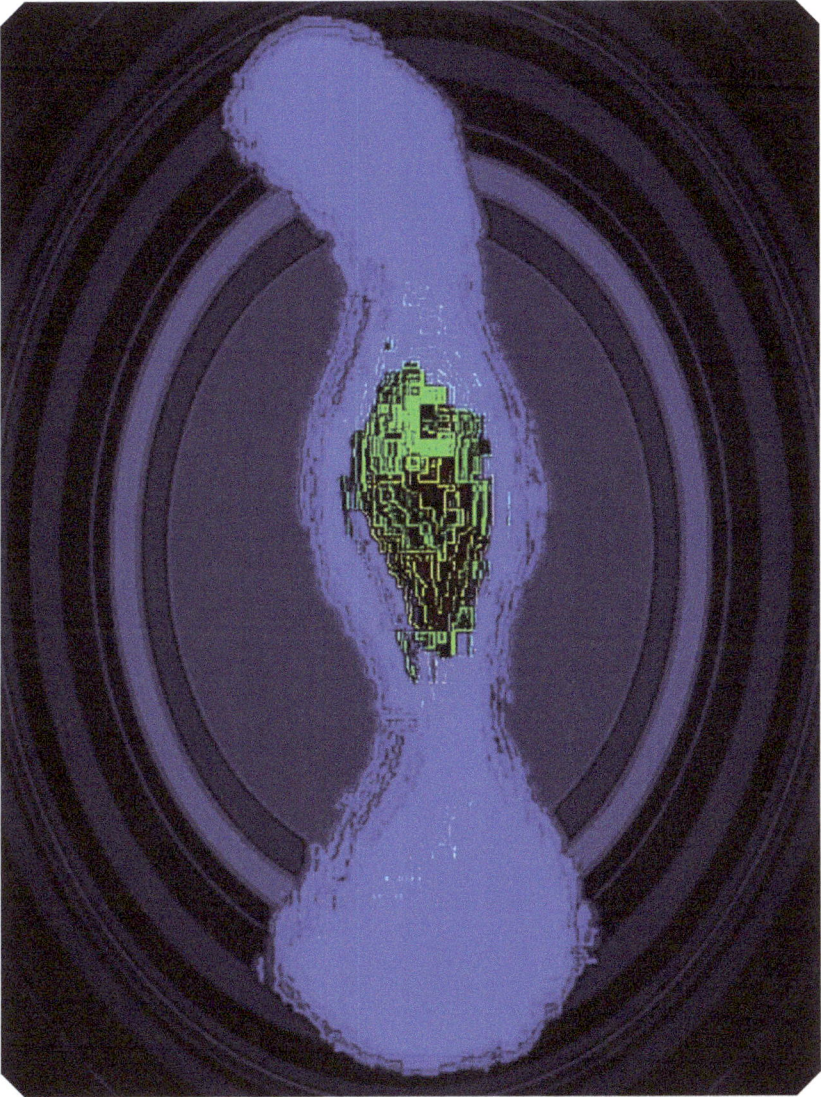

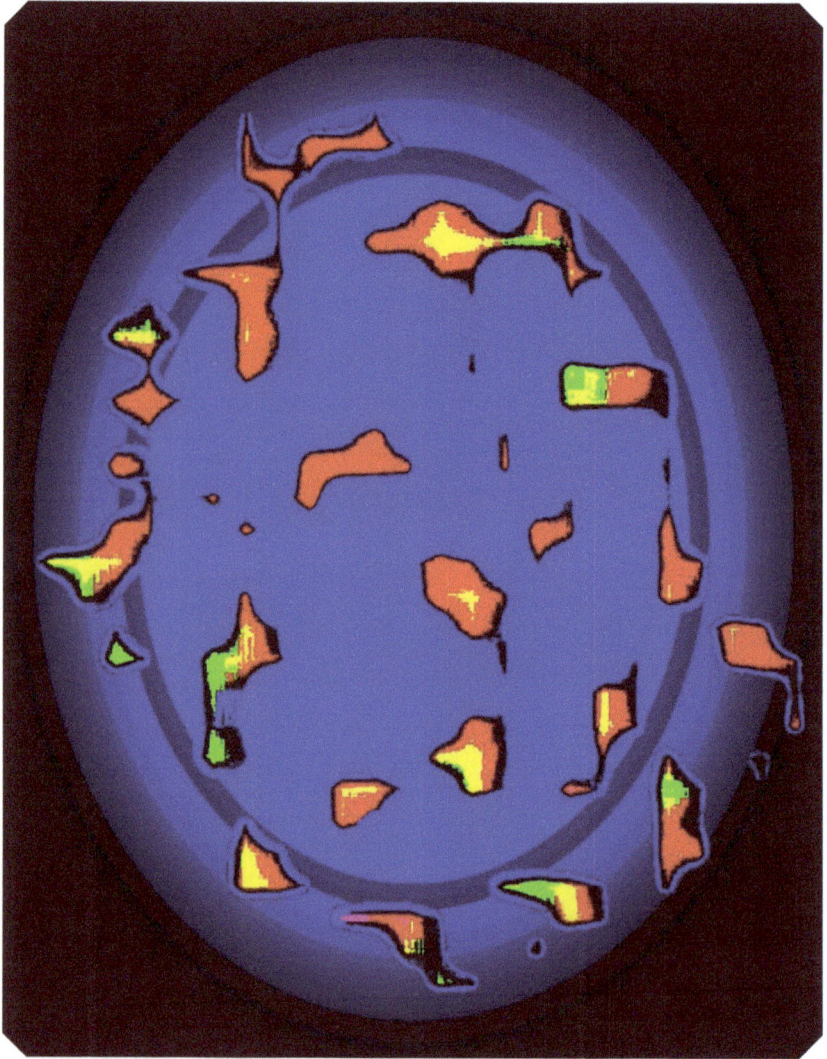

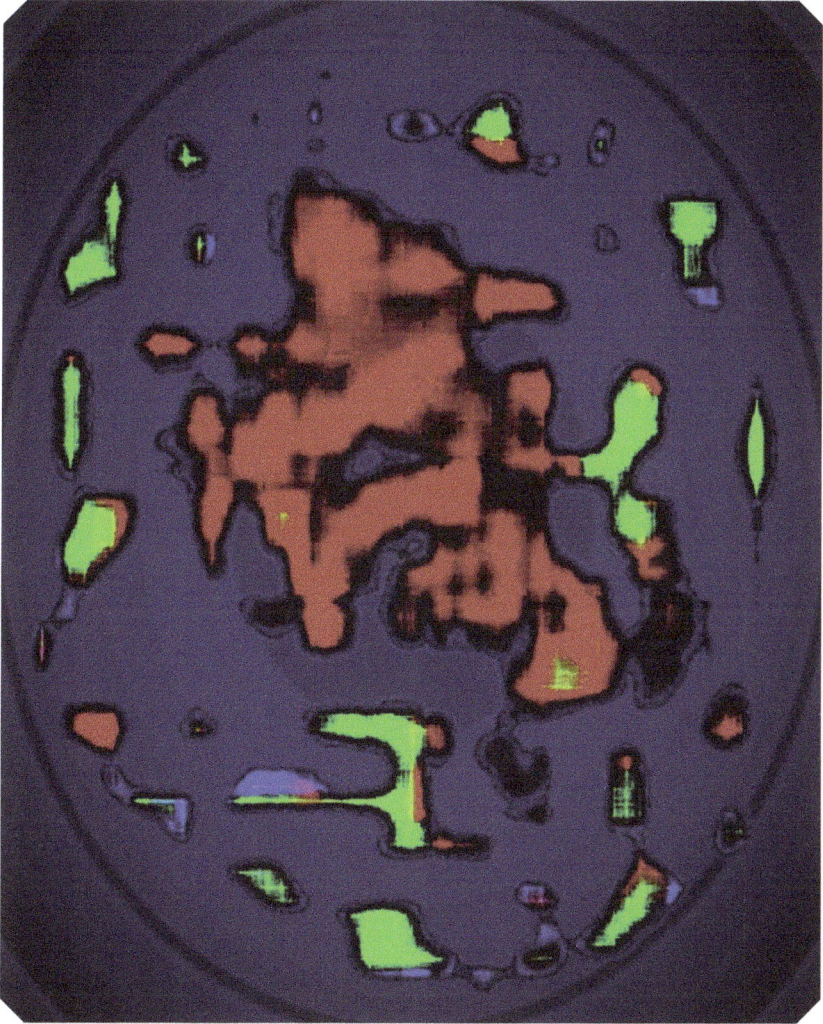

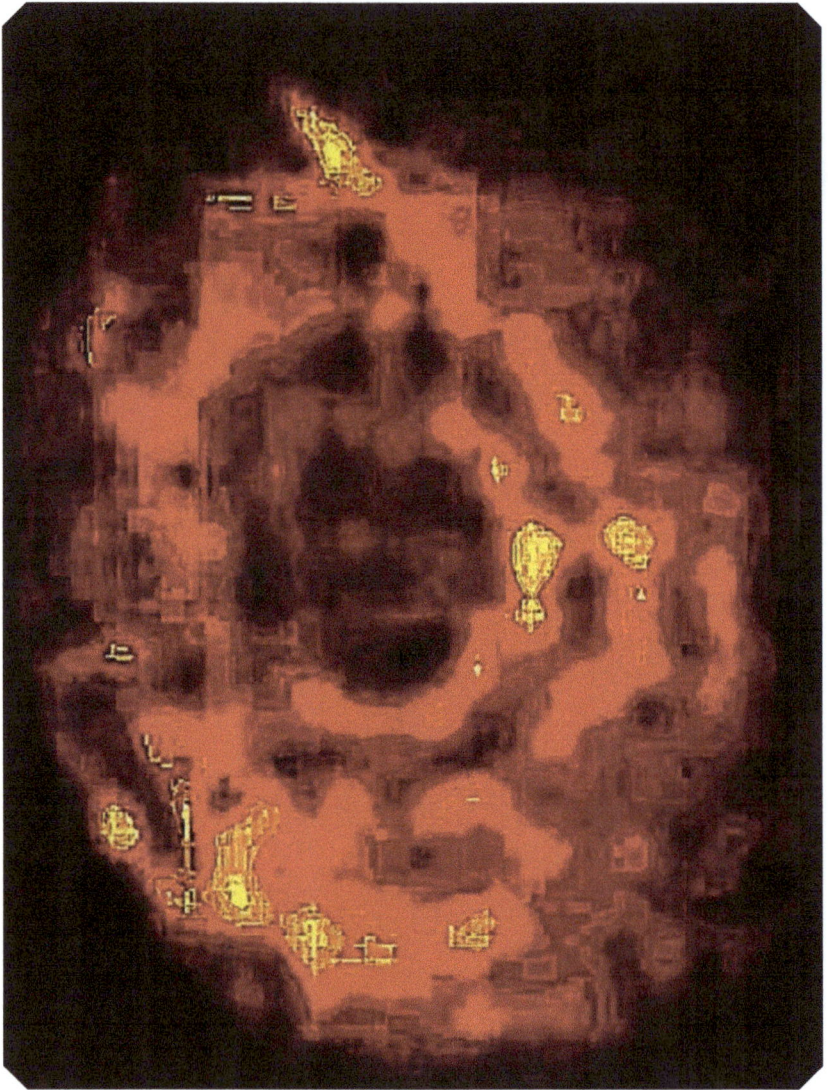

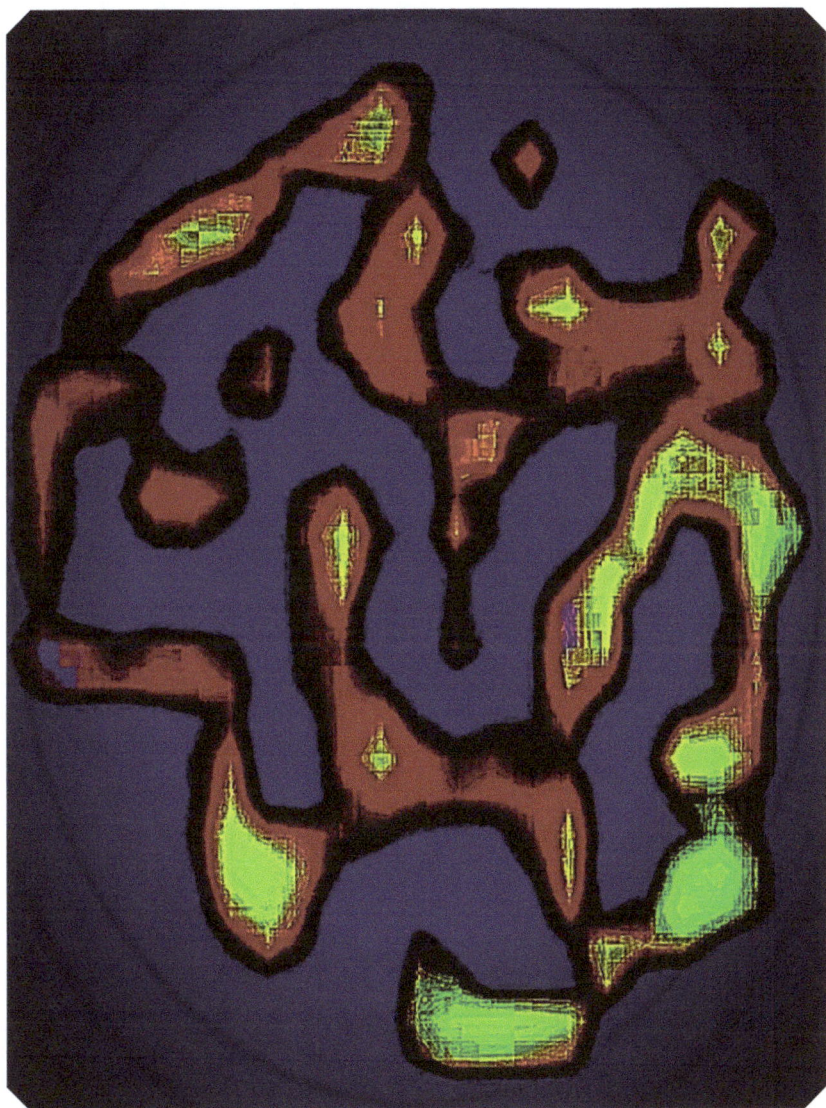

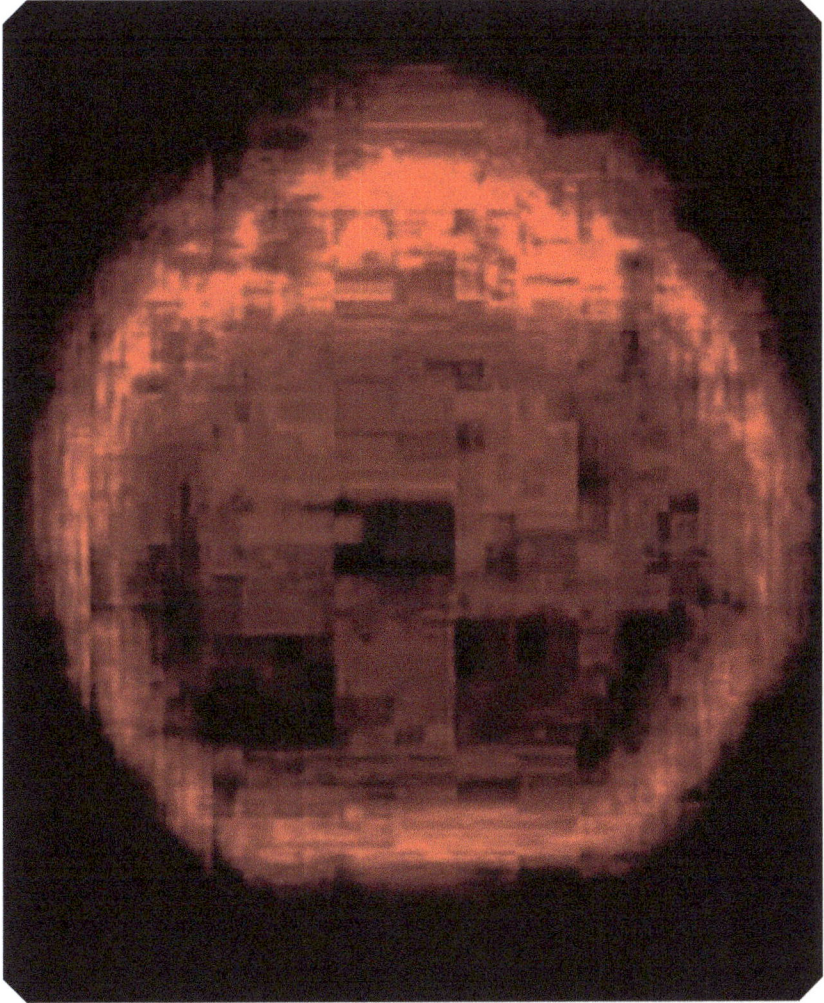

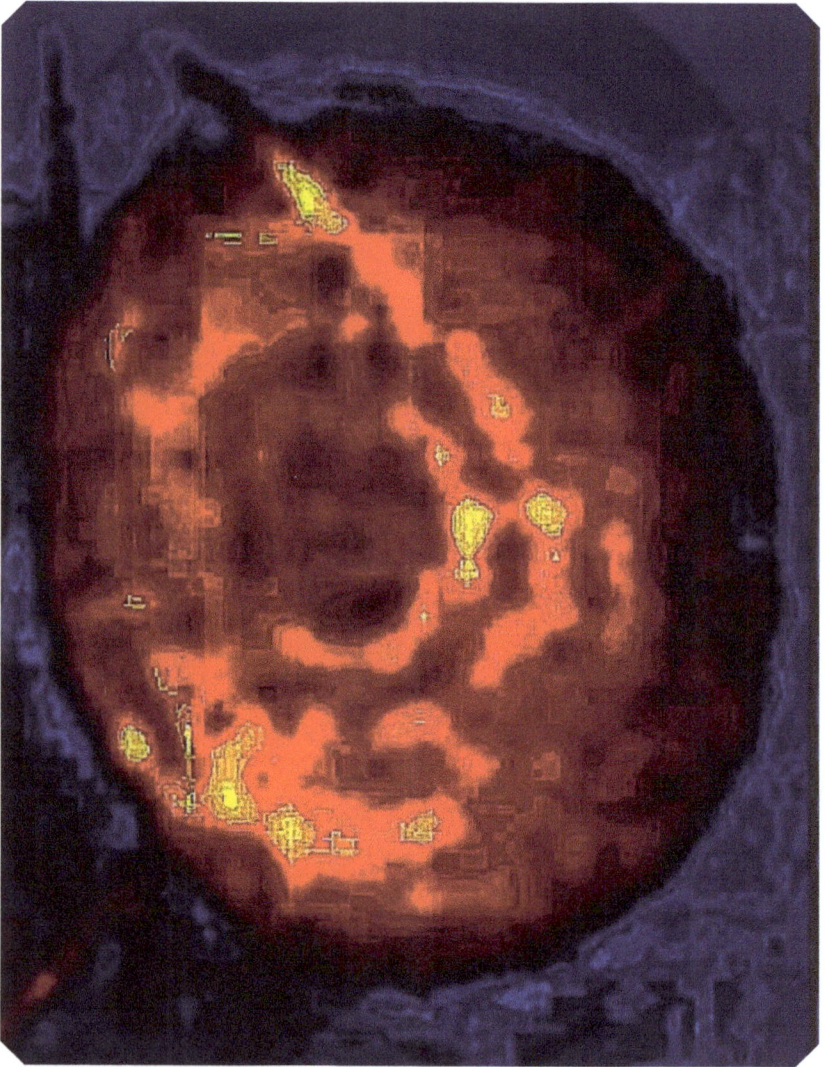

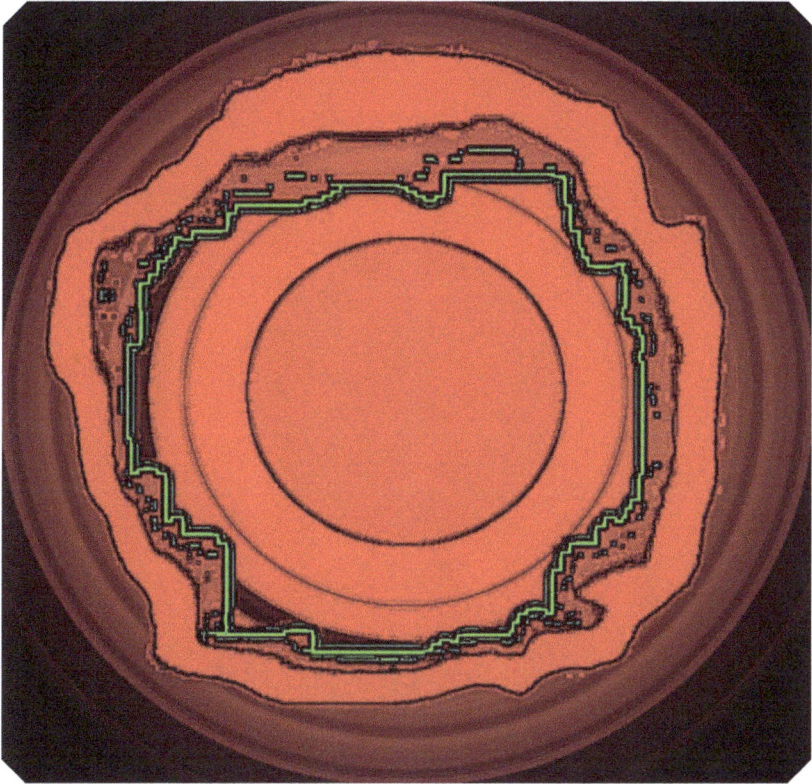

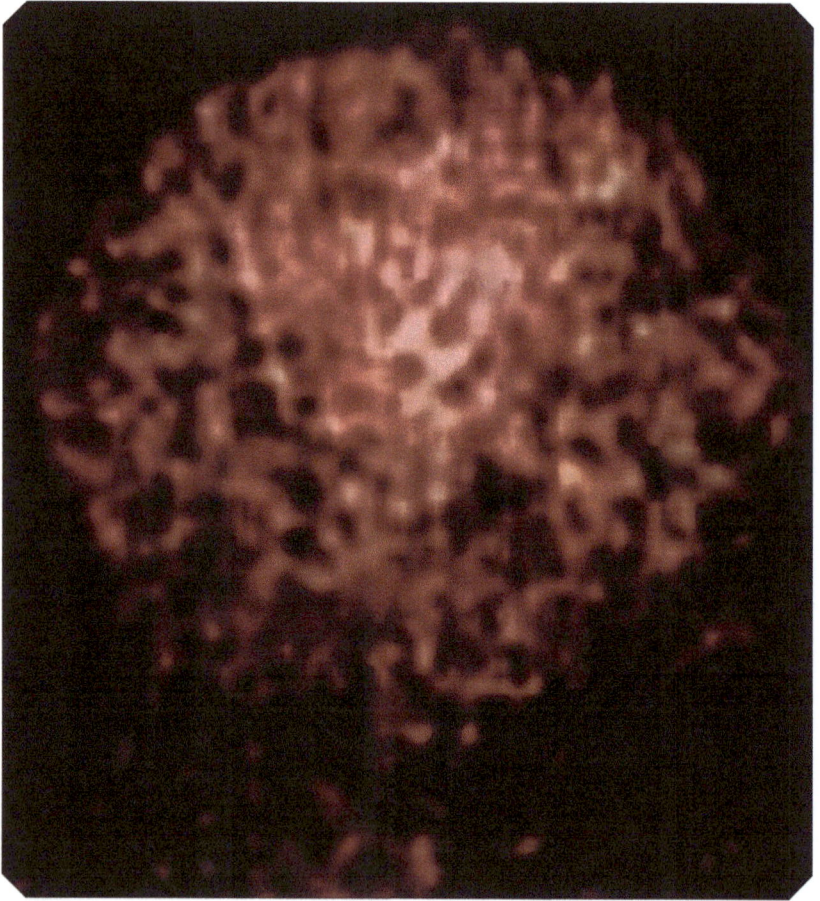

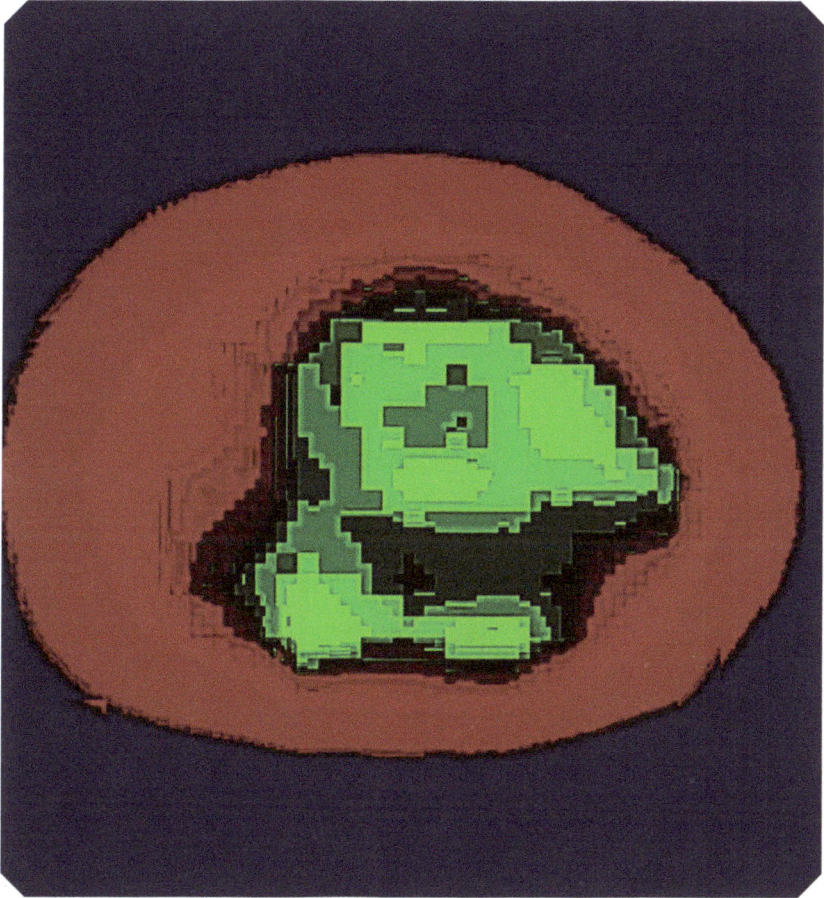

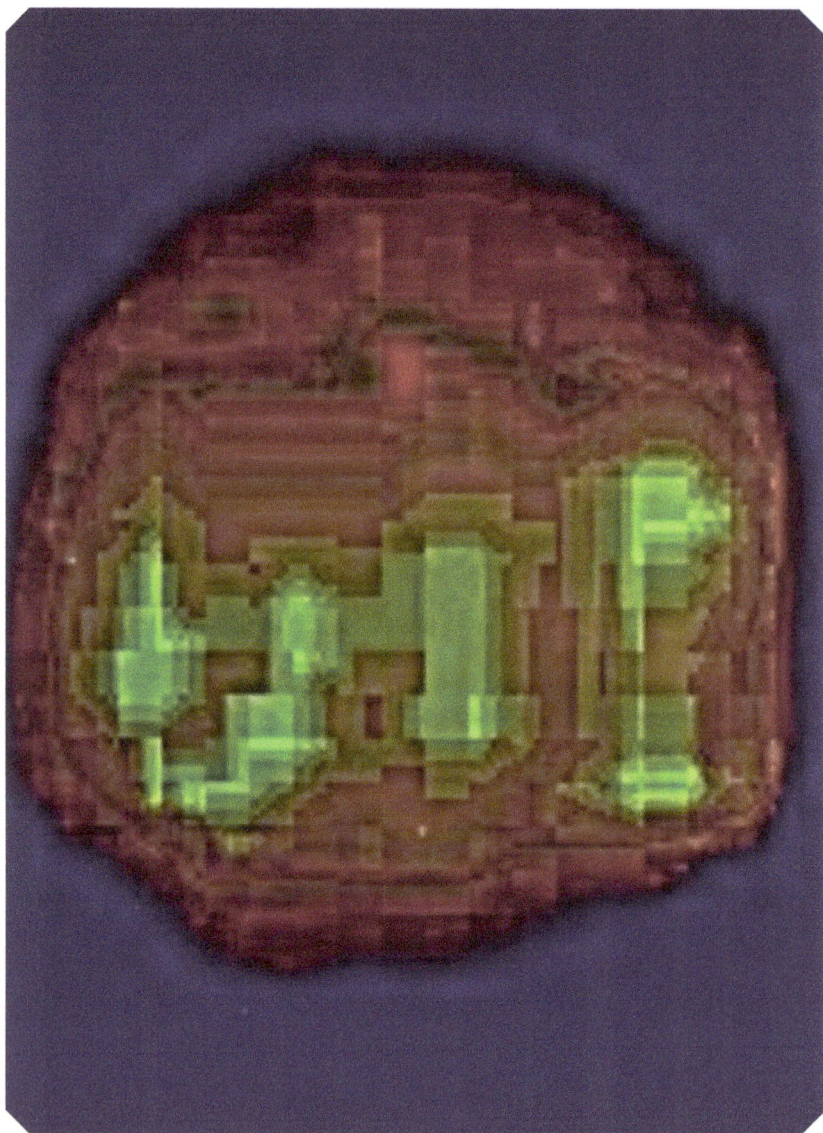

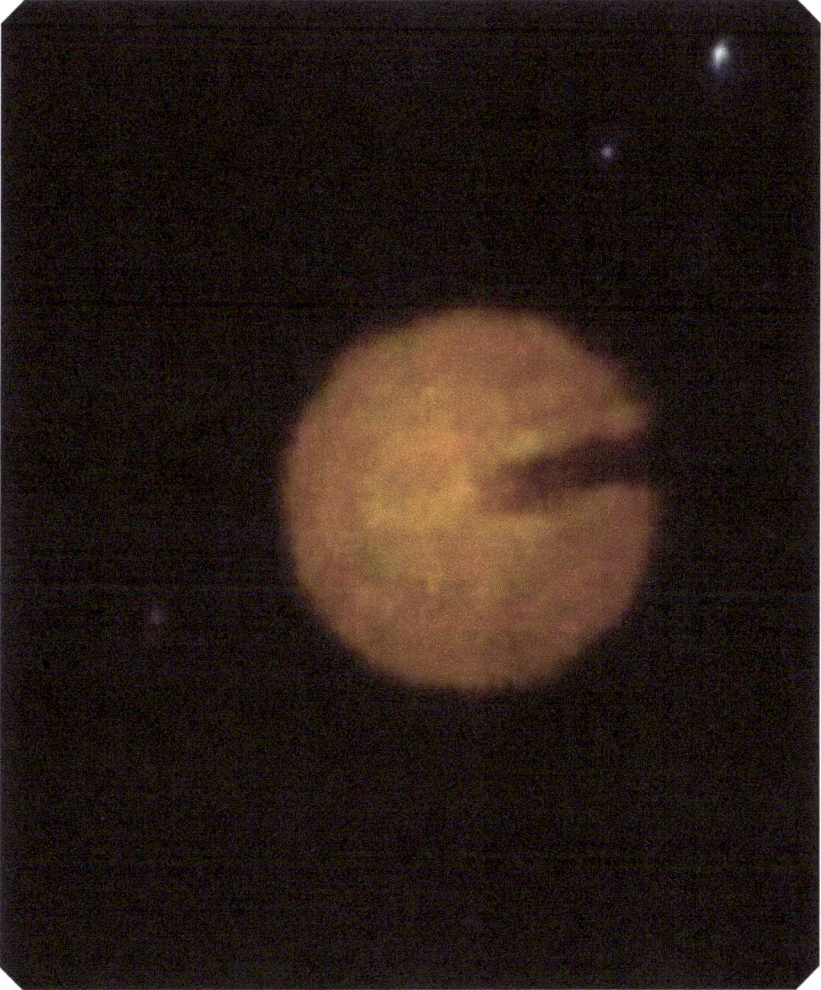

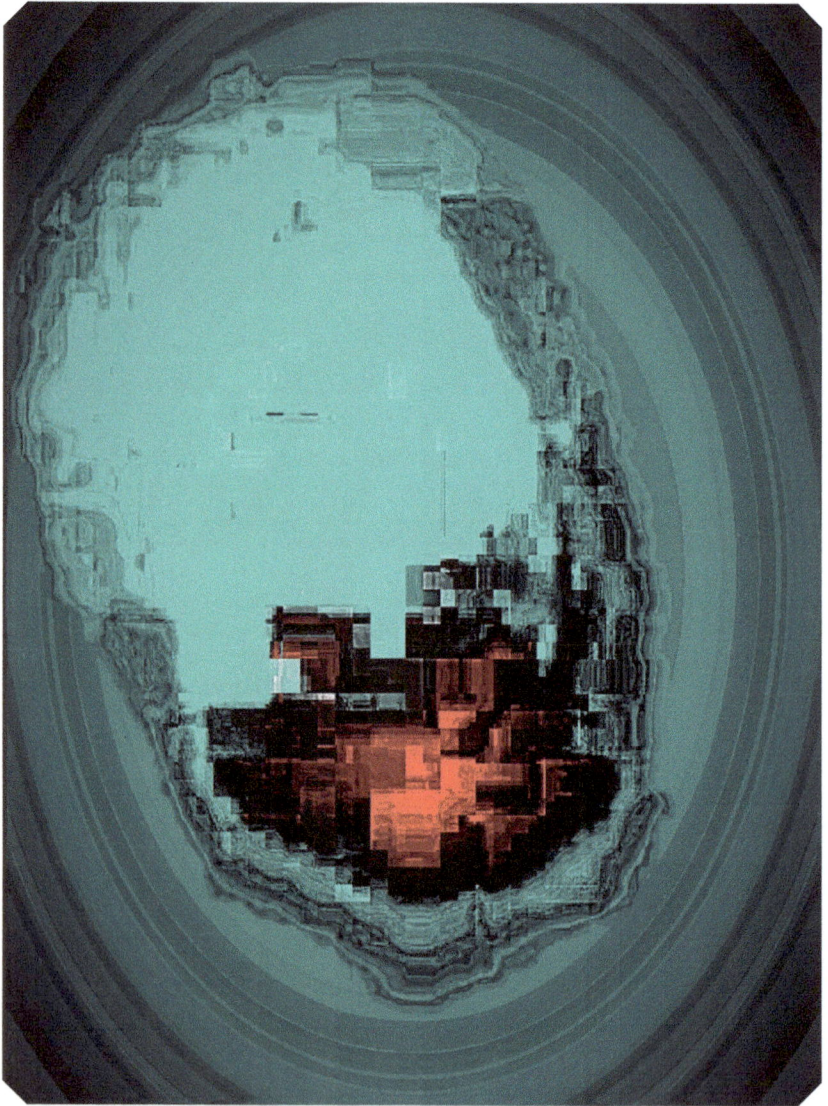

CATEGORY TWO

RECTANGULAR STARSHIPS

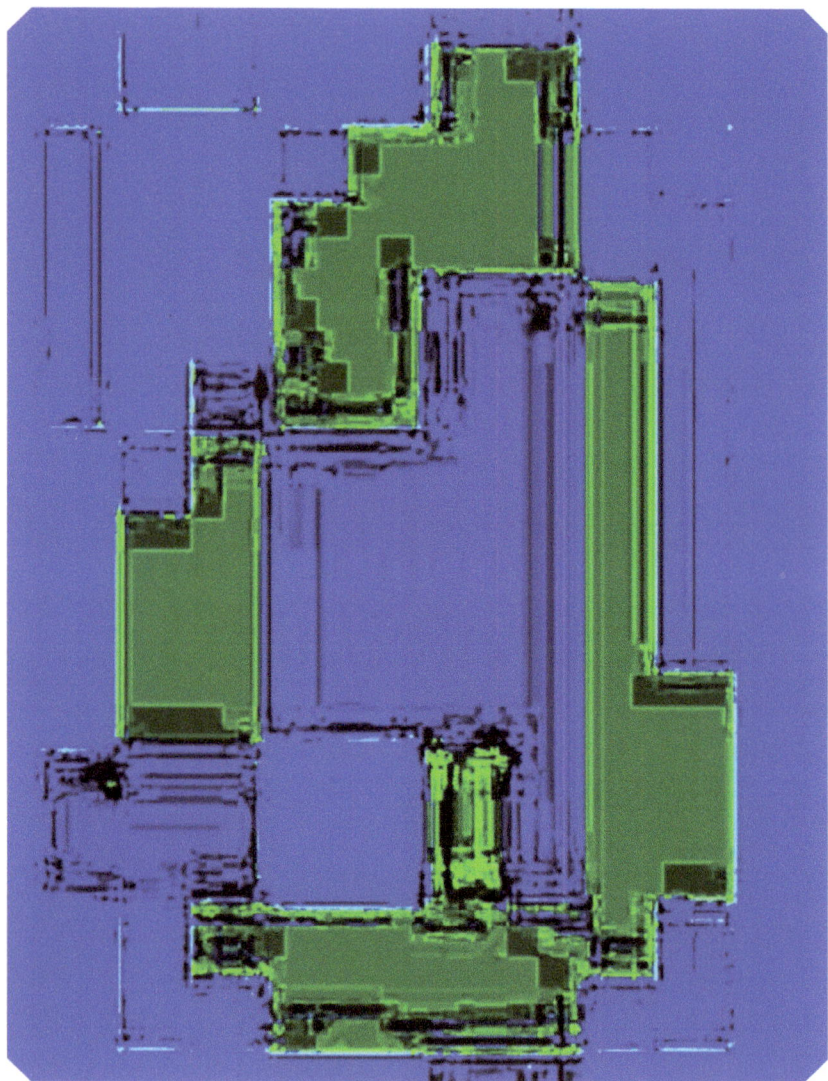

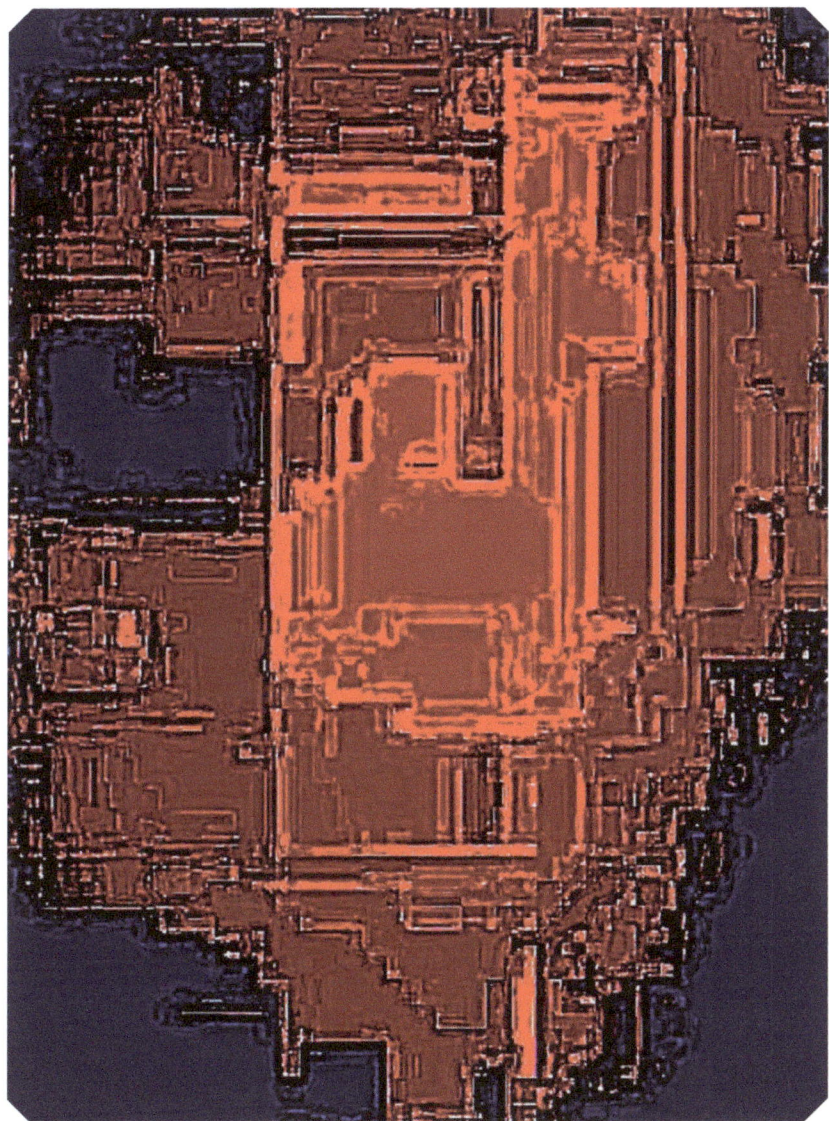

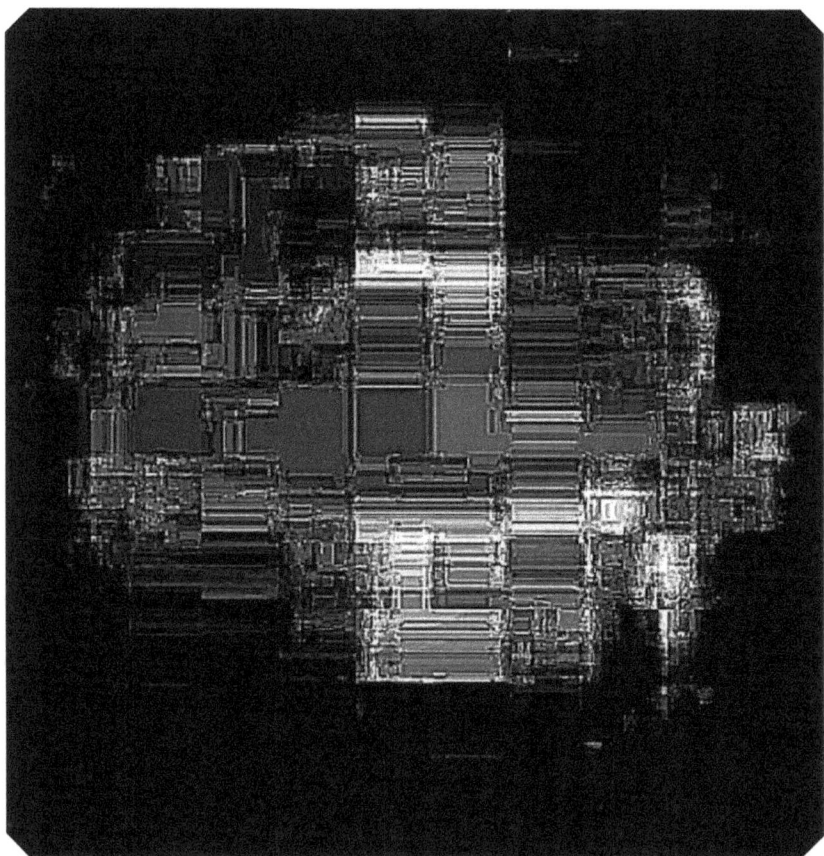

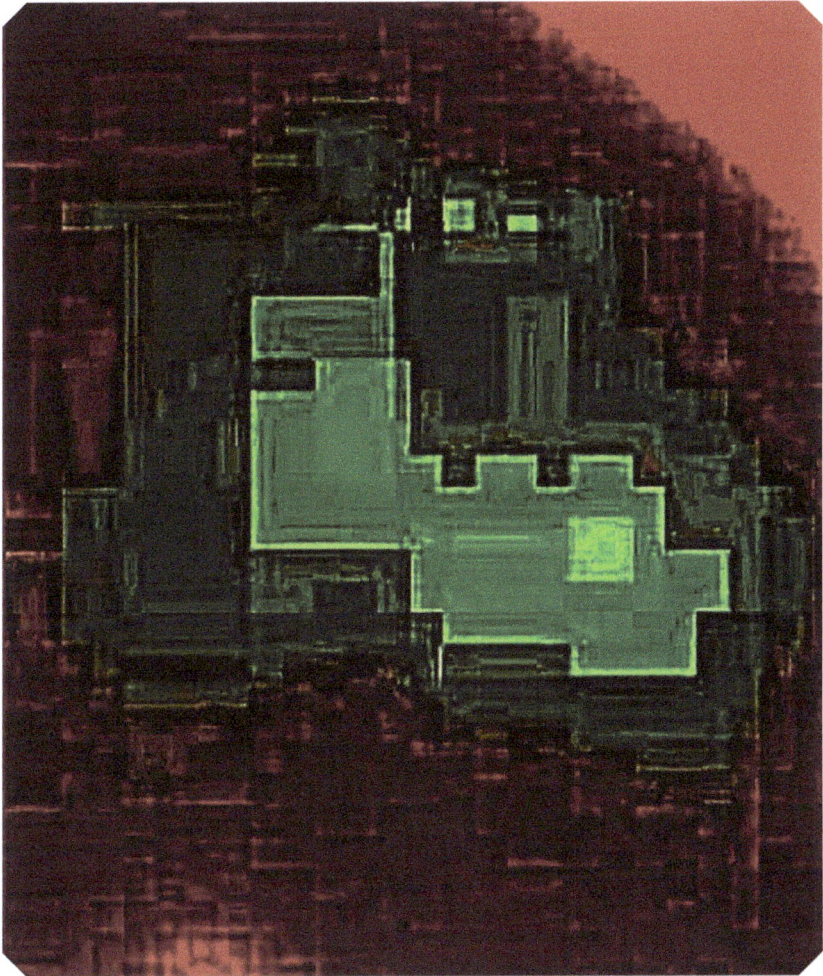

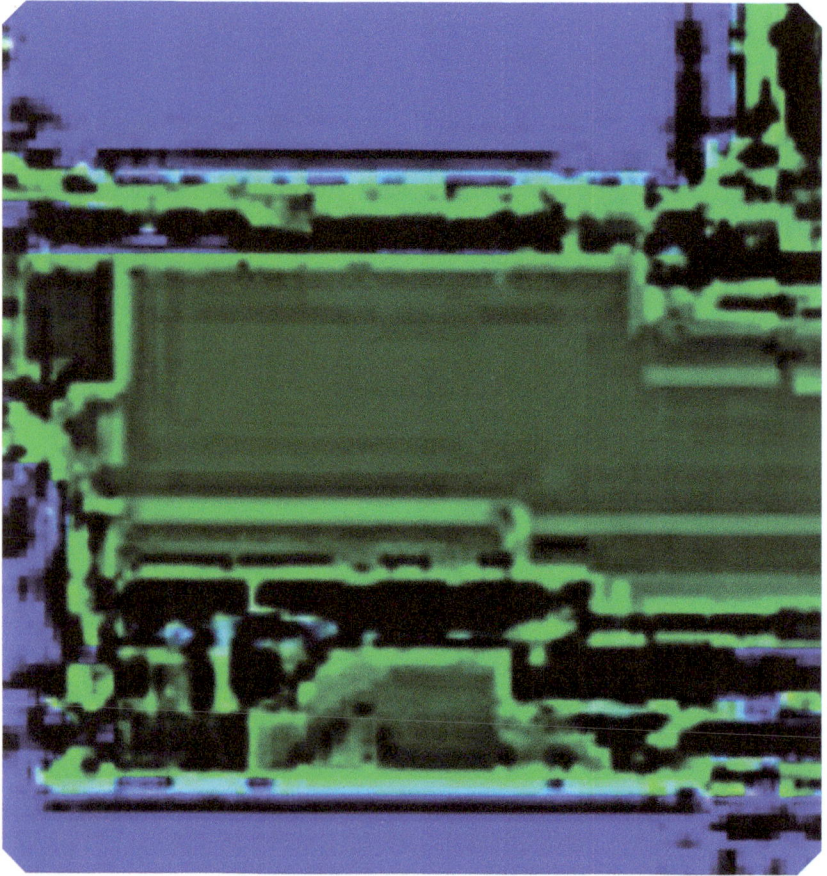

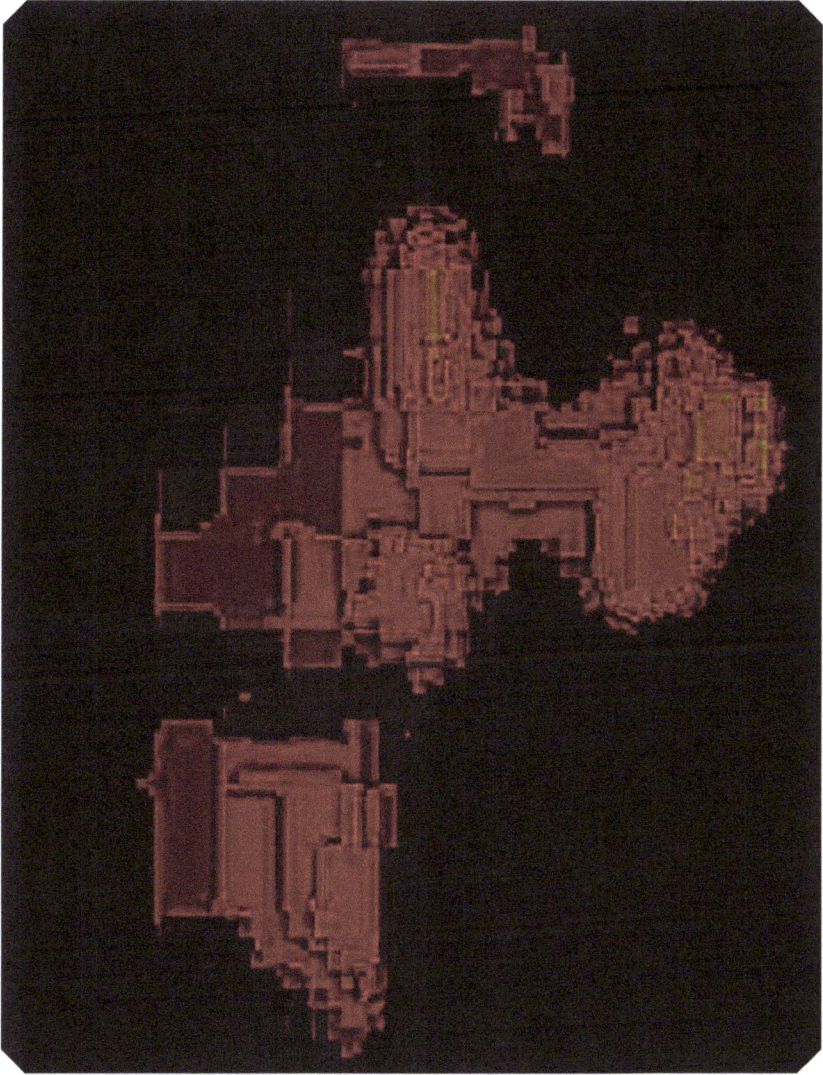

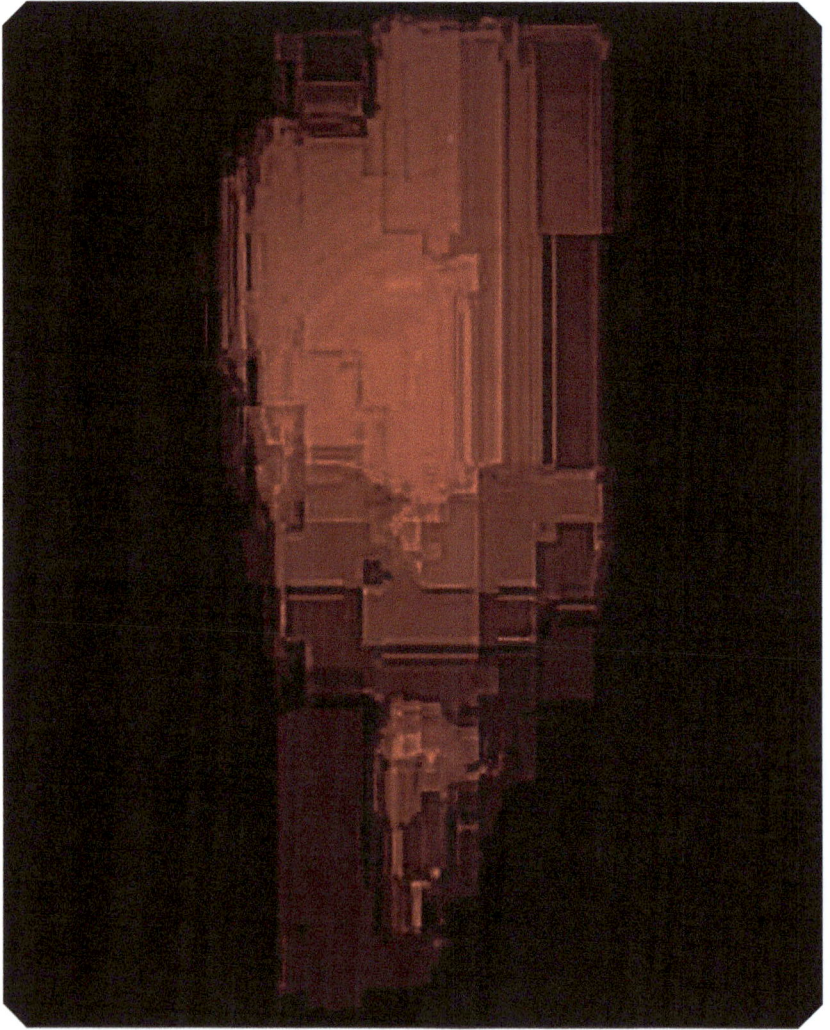

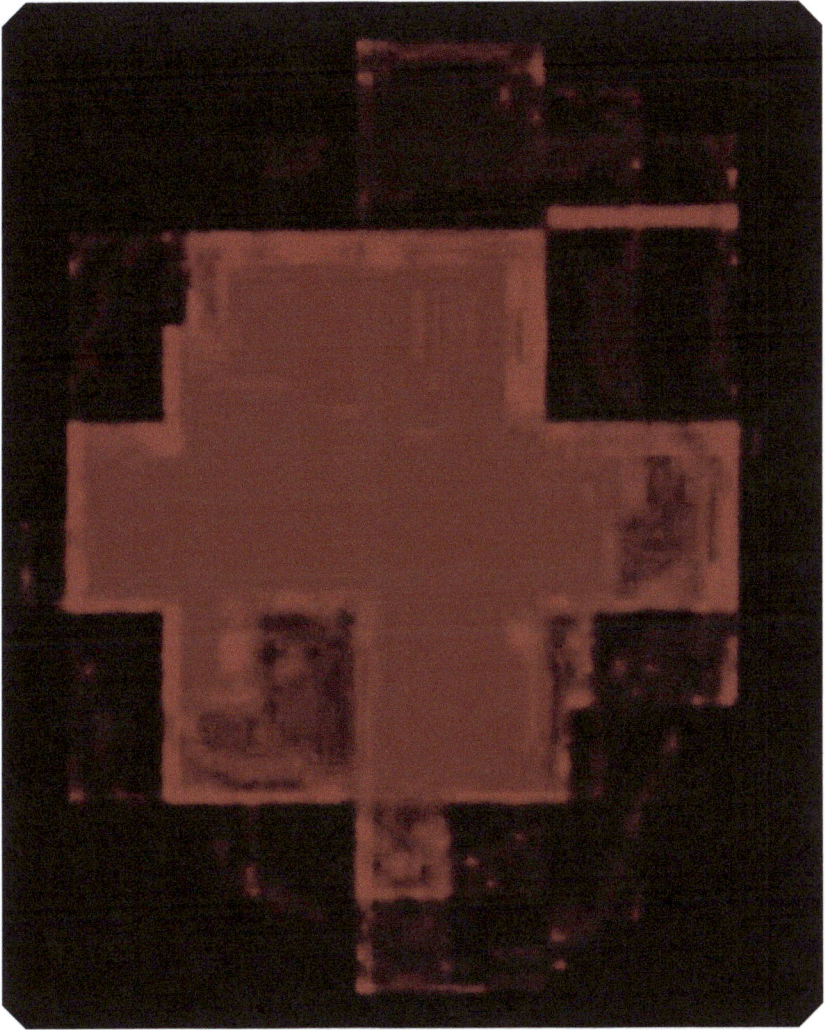

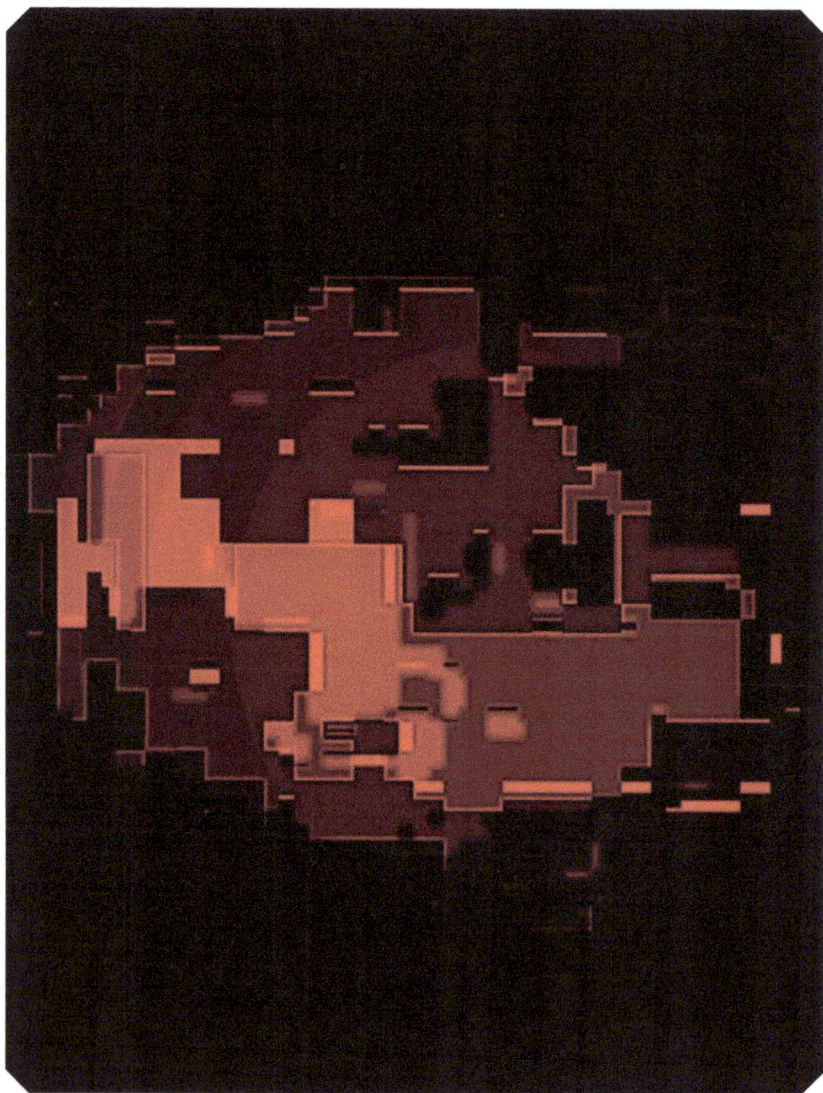

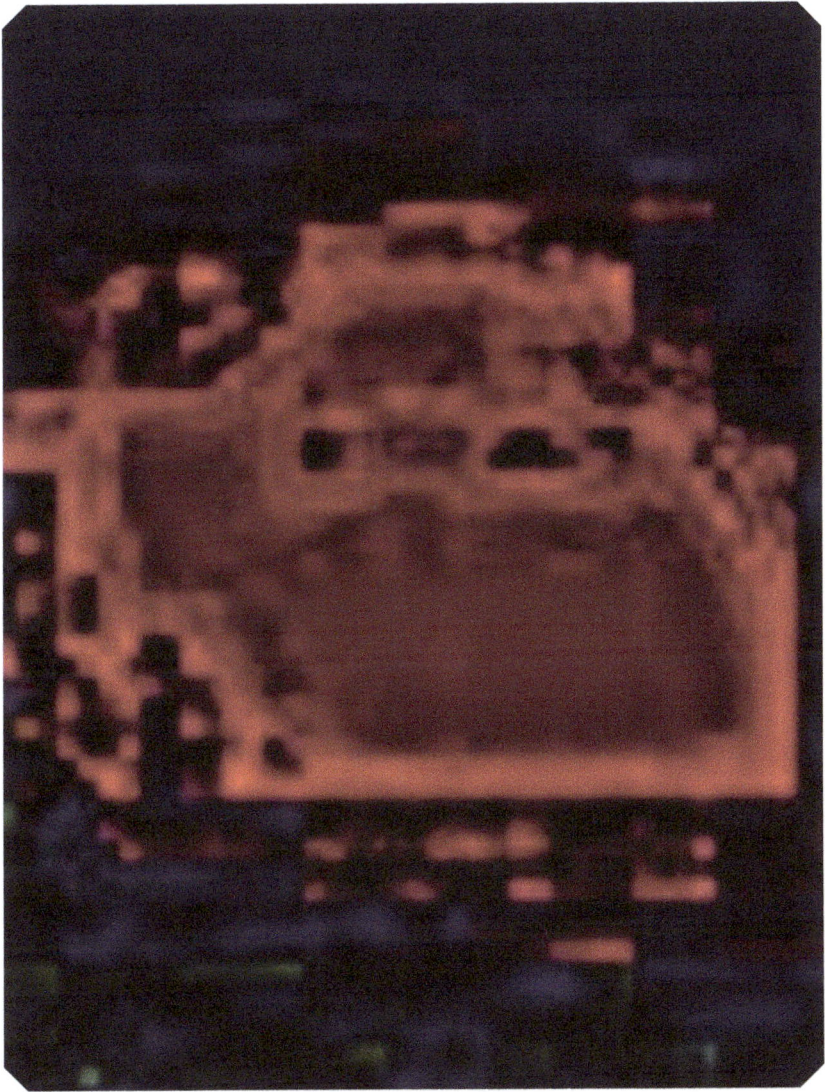

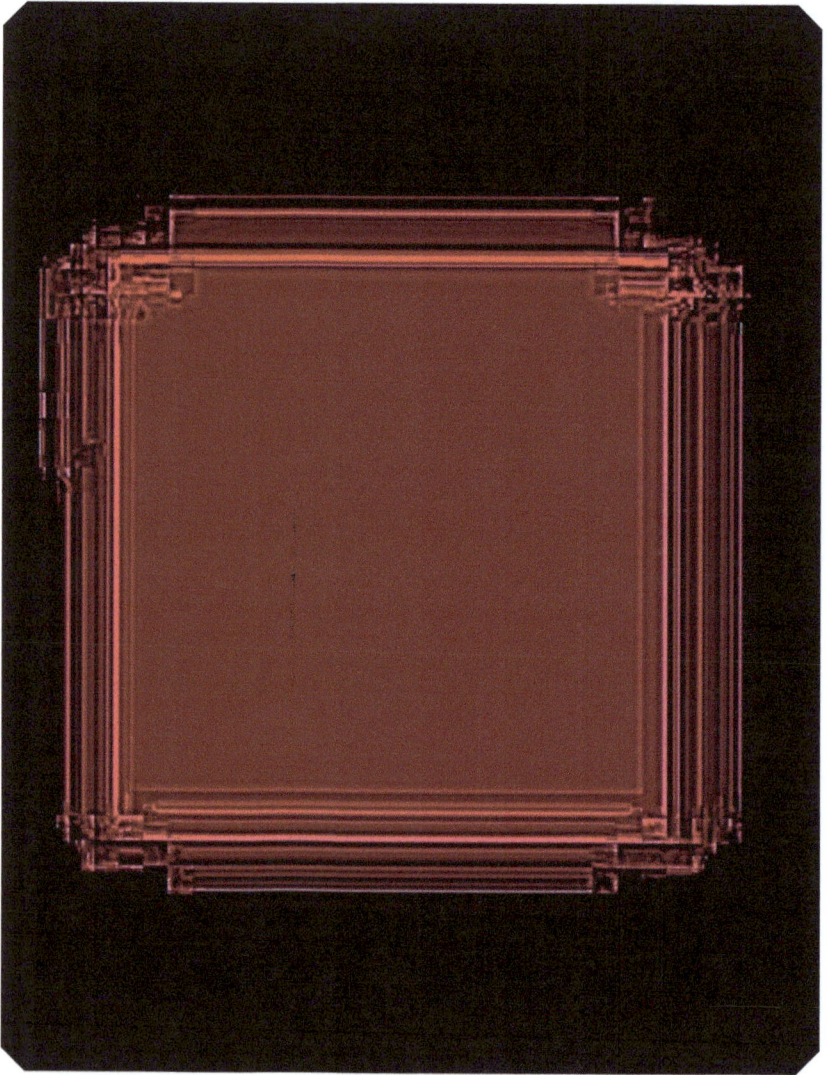

CATEGORY THREE

ANOMALOUS
STARSHIPS

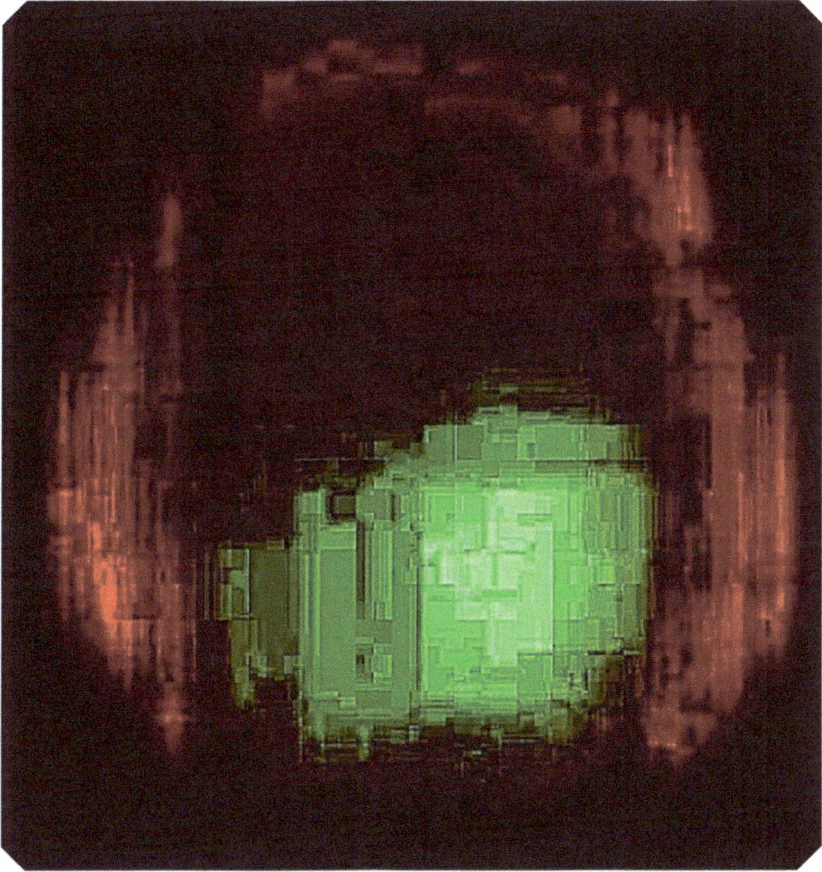

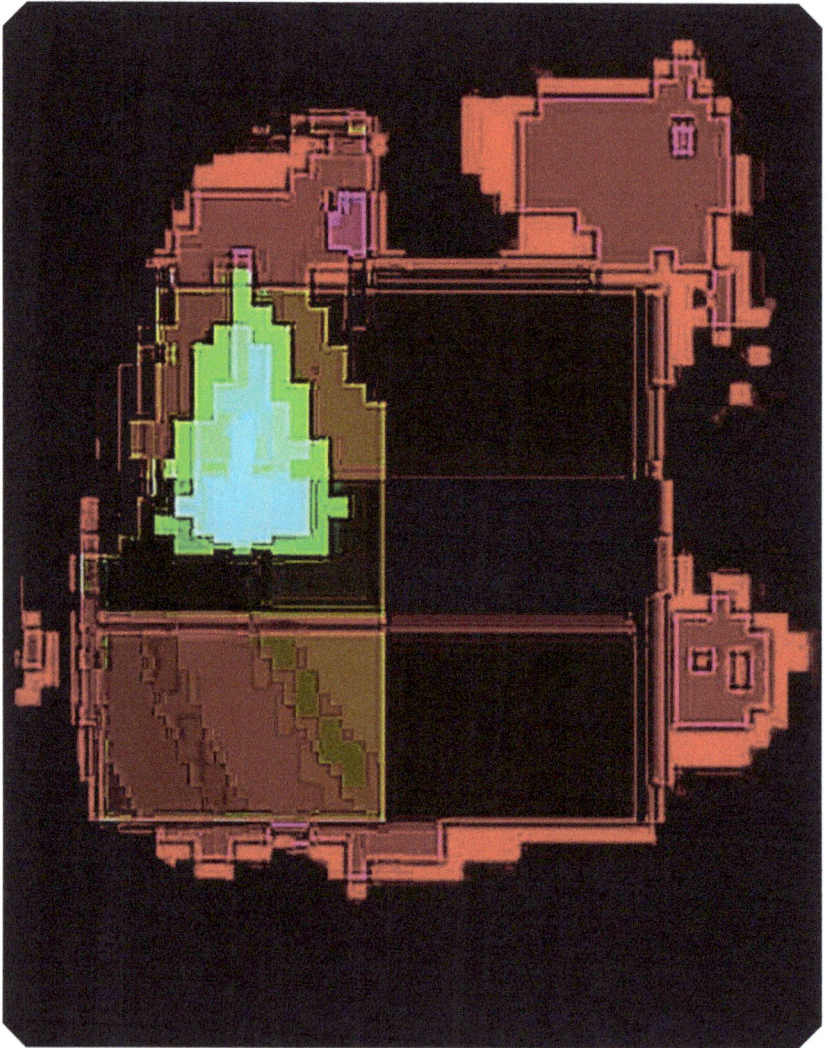

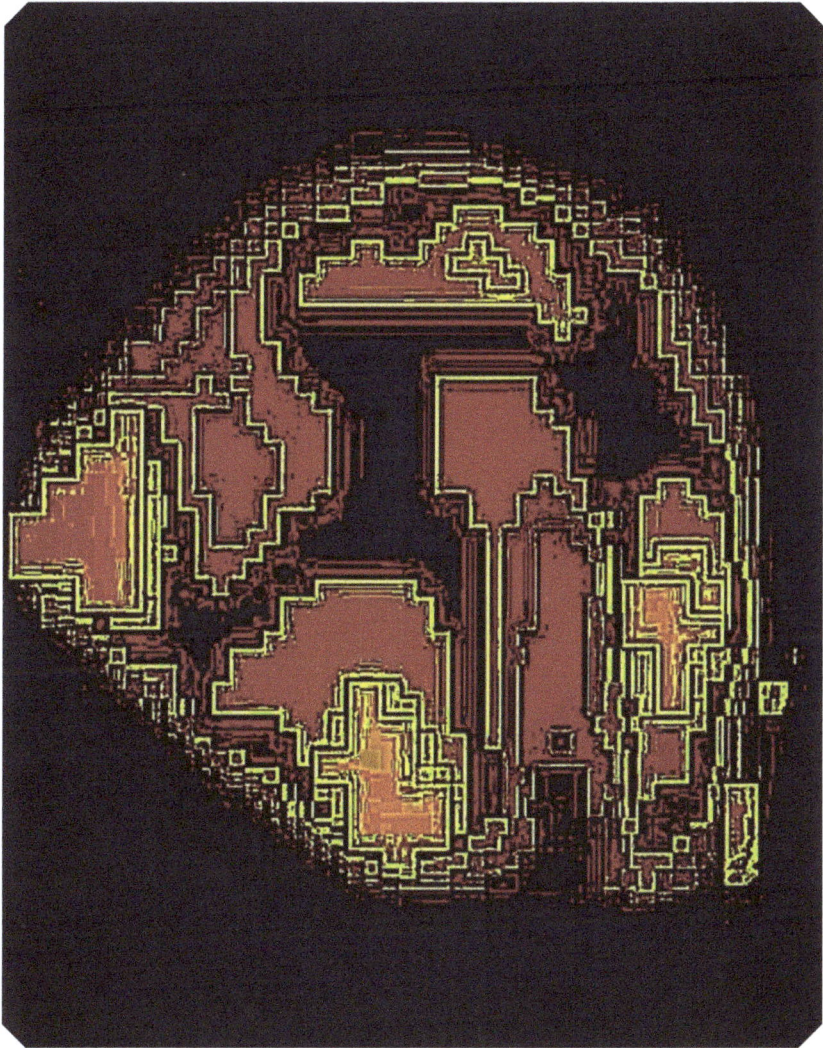

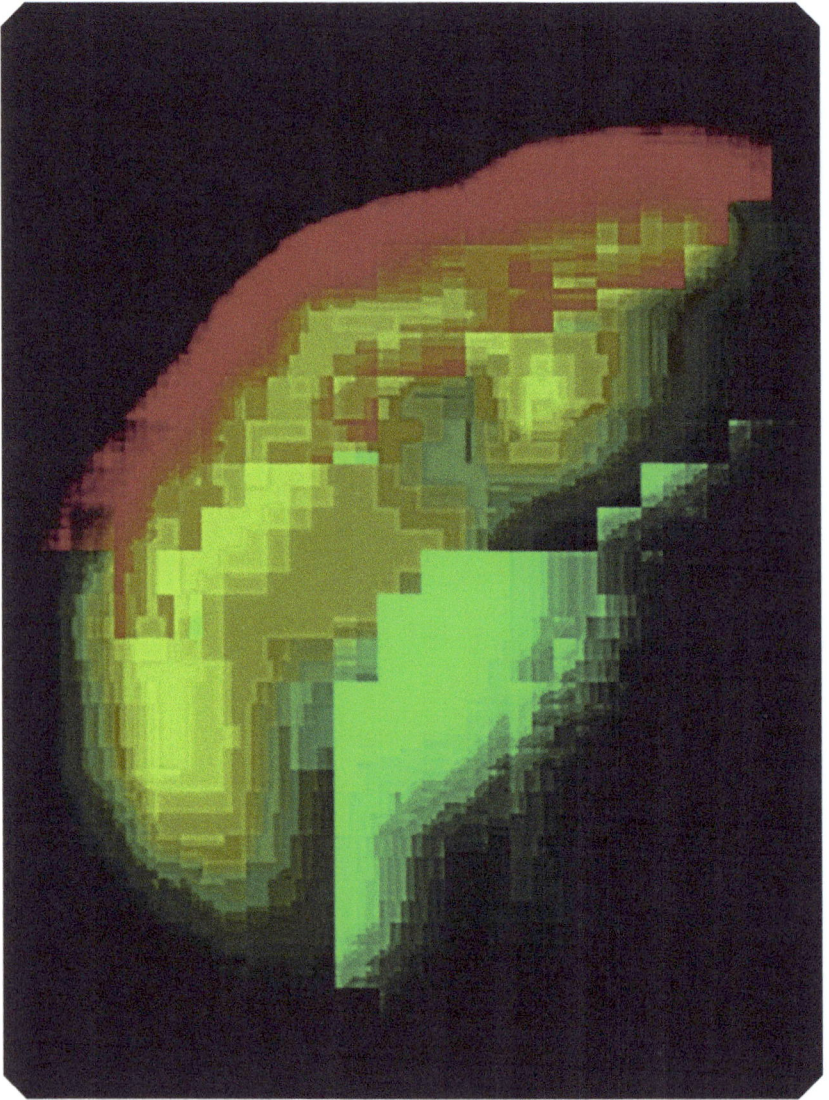

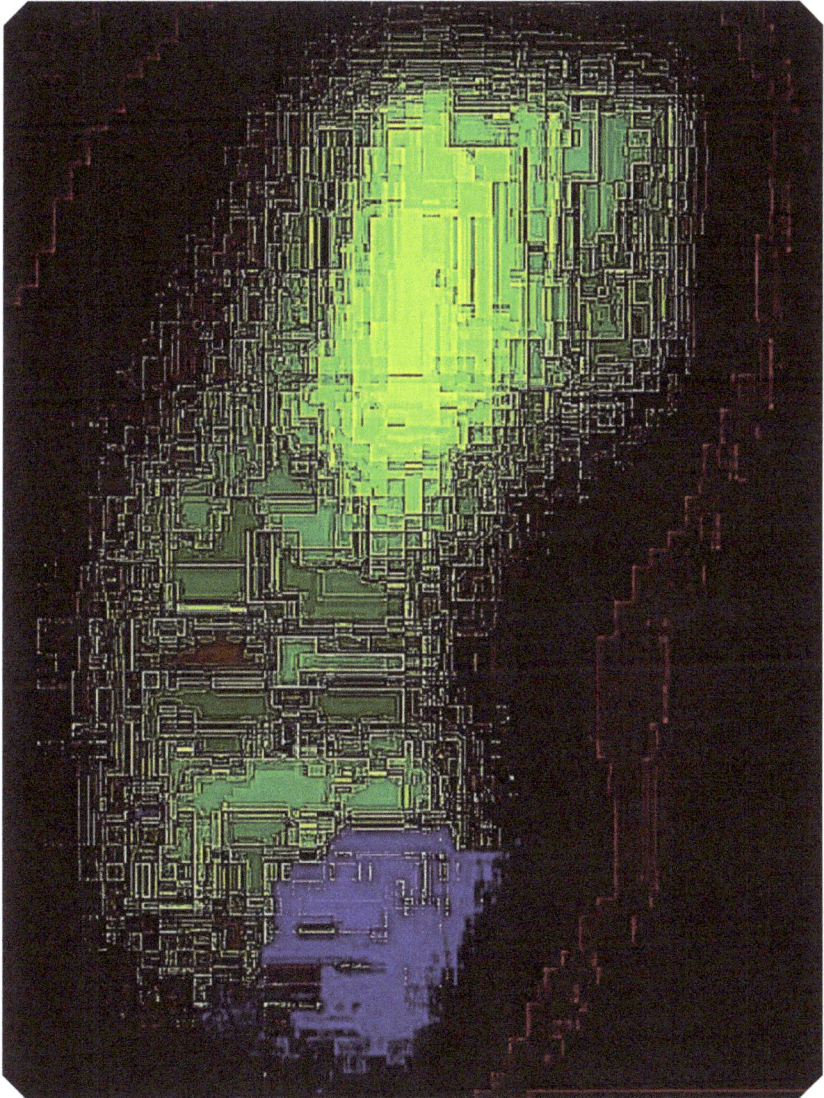

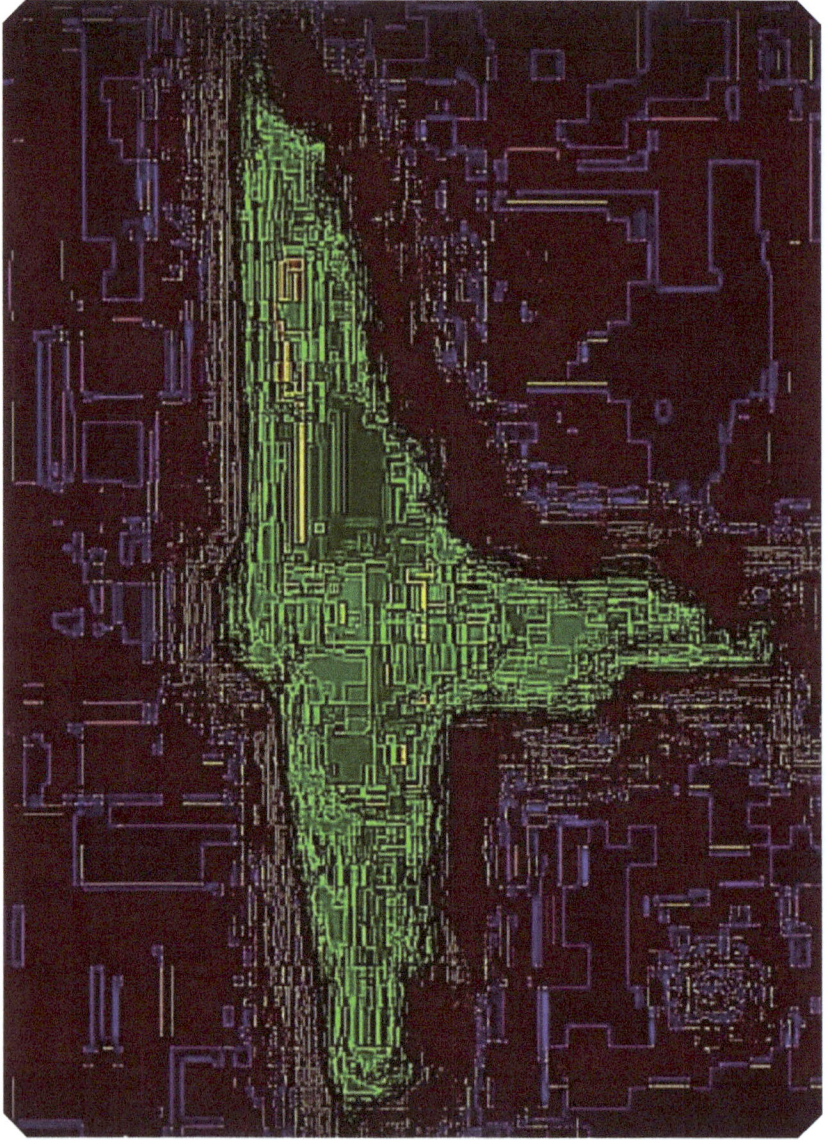

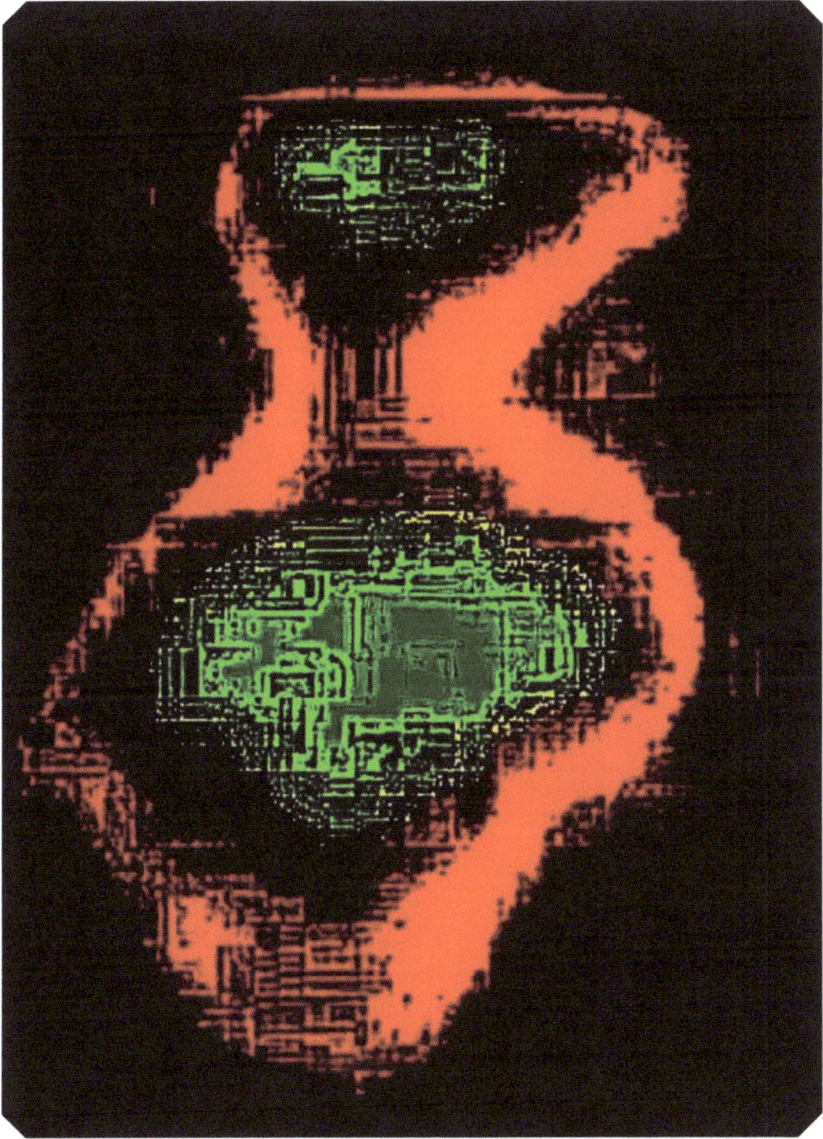

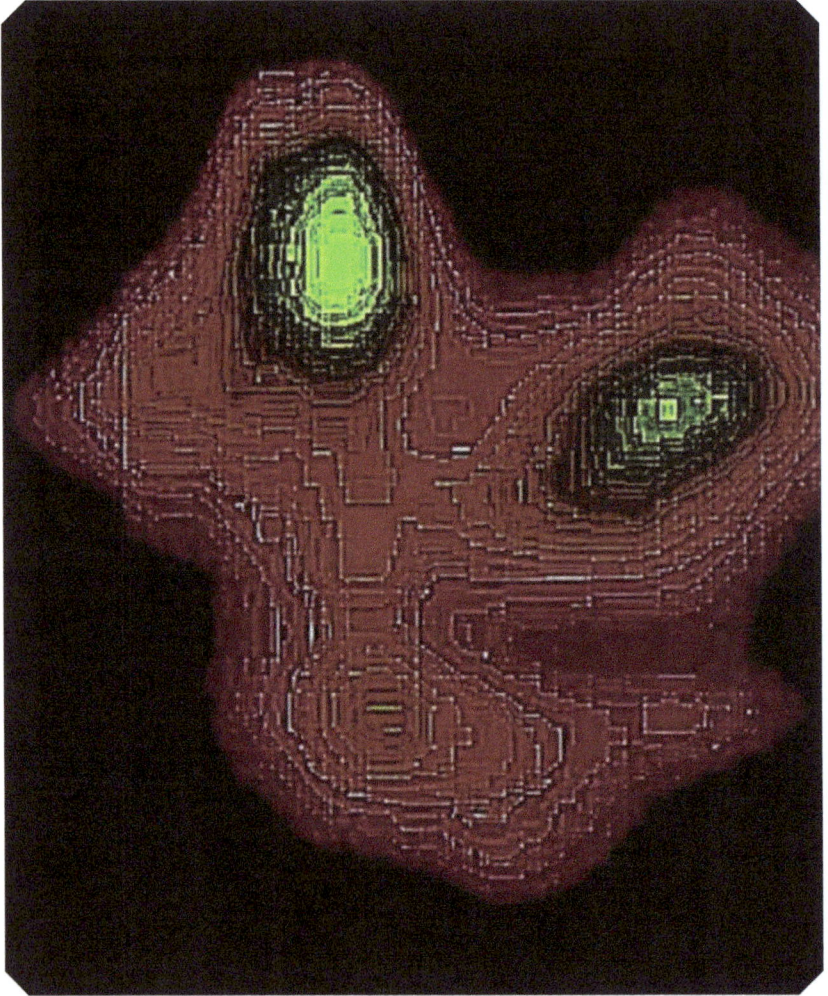

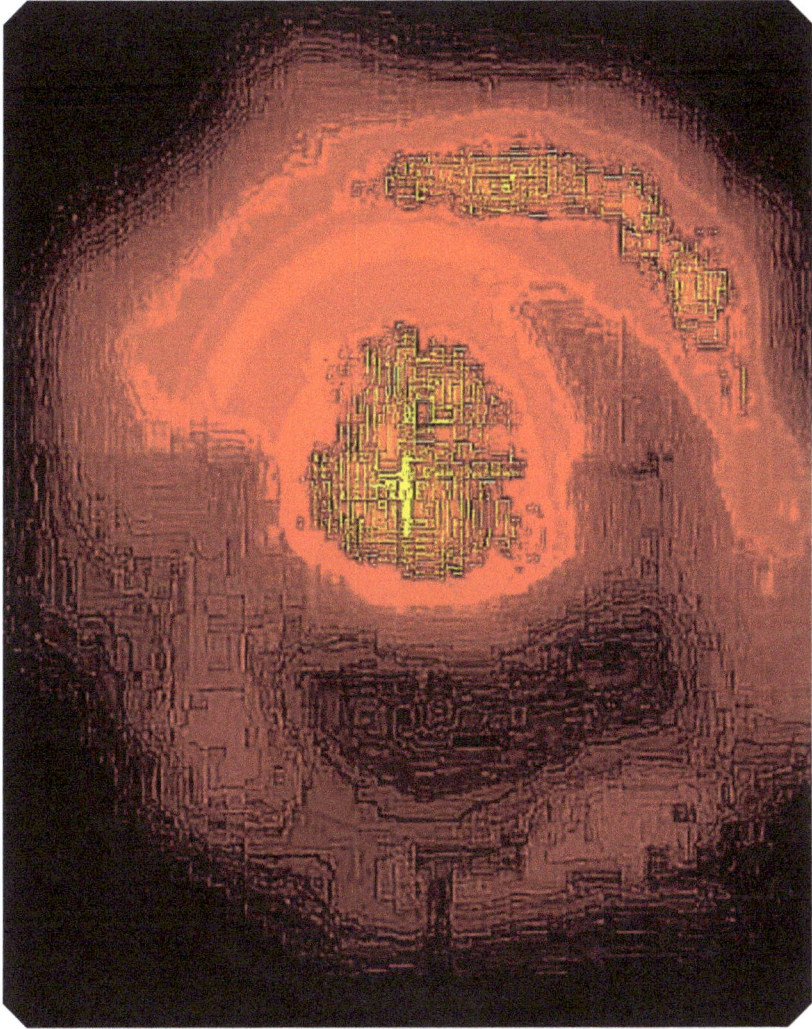

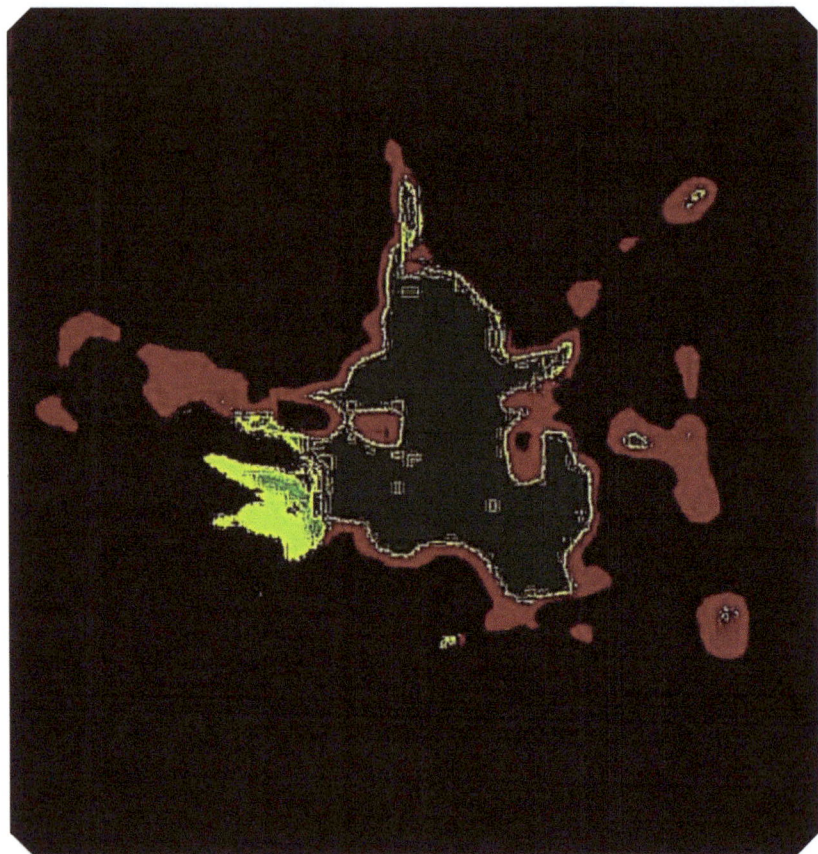

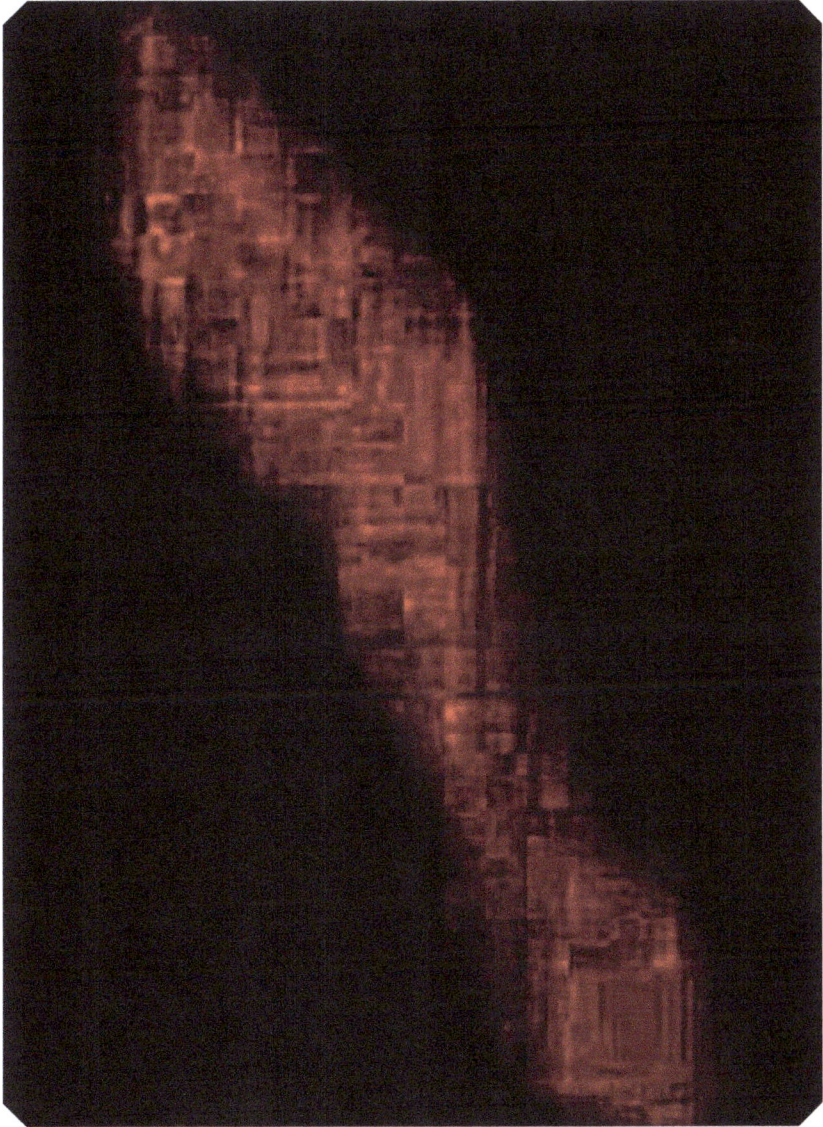

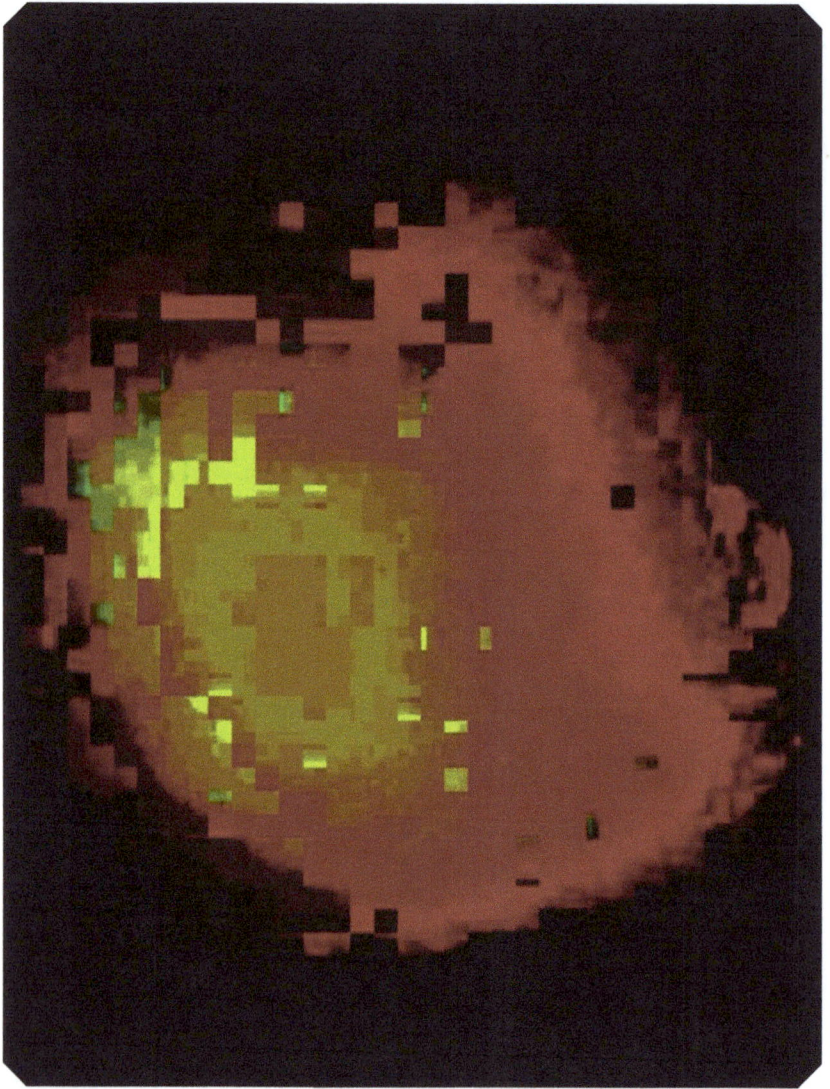

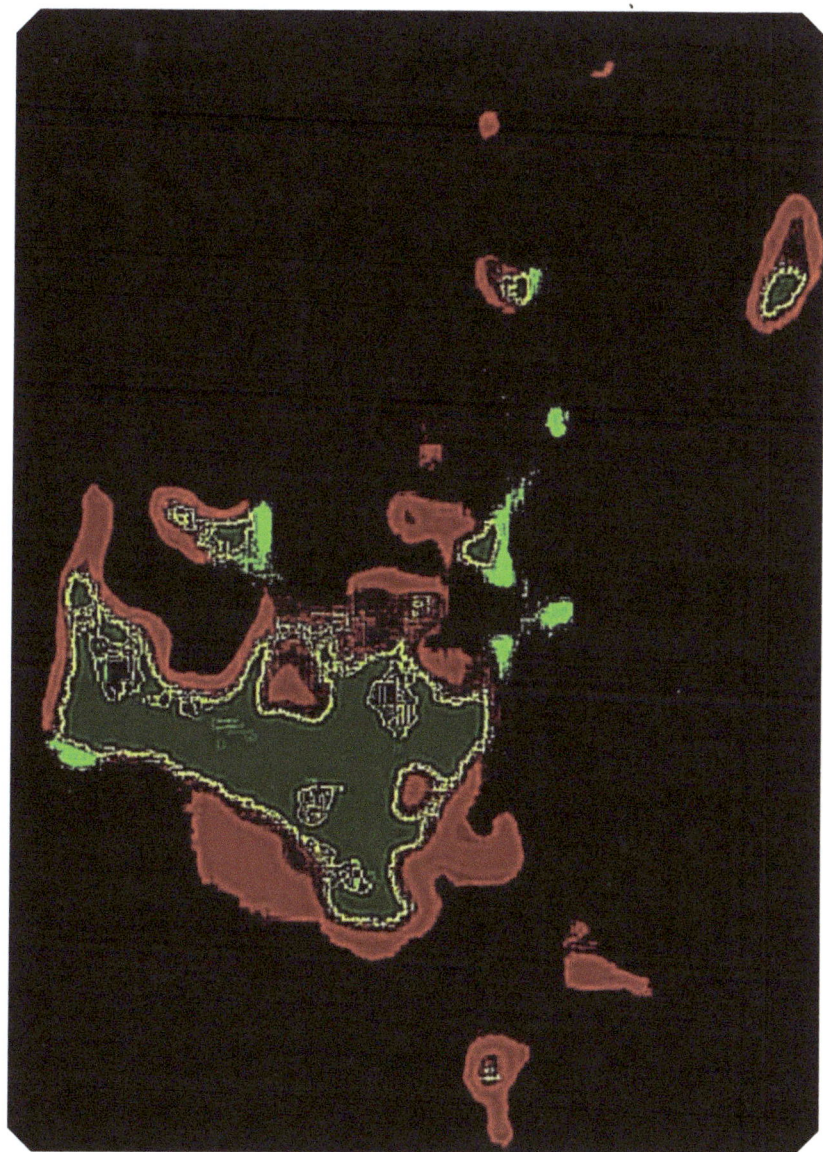

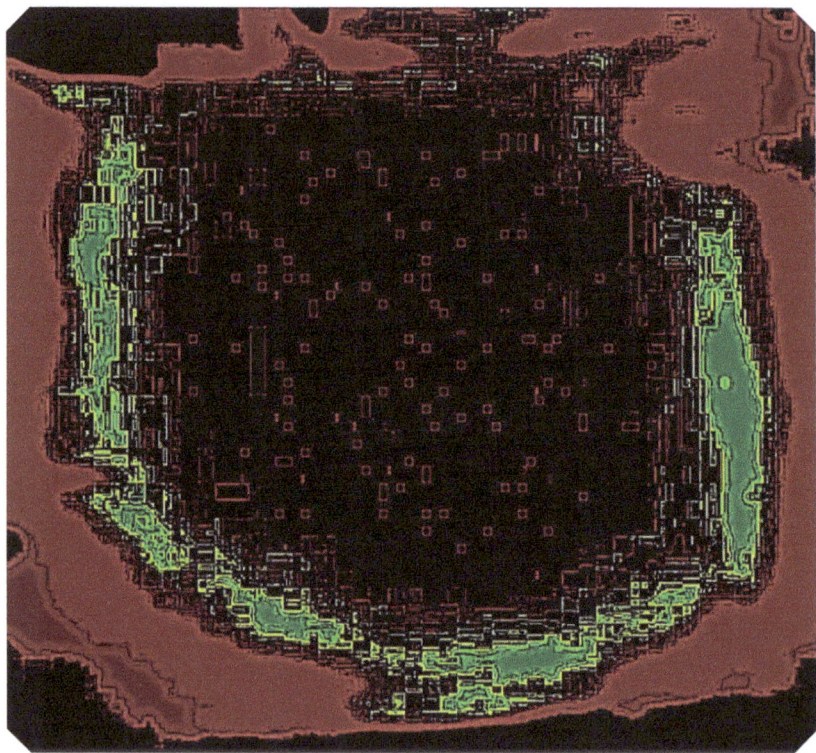

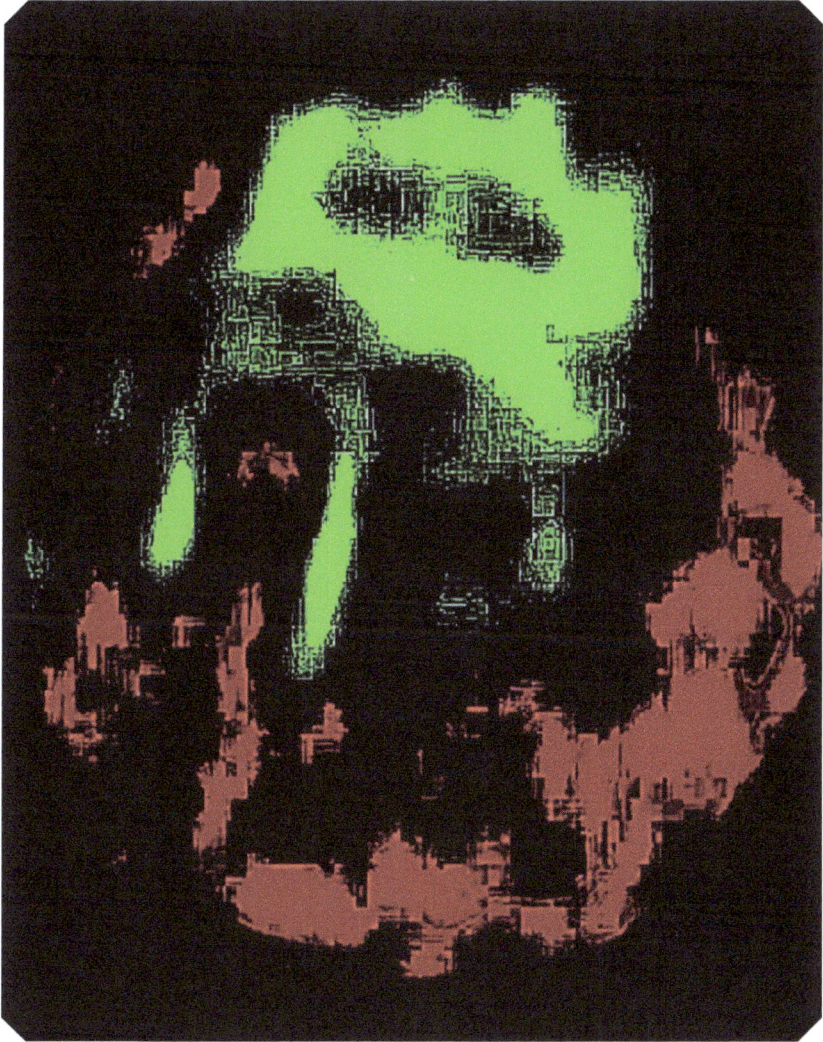

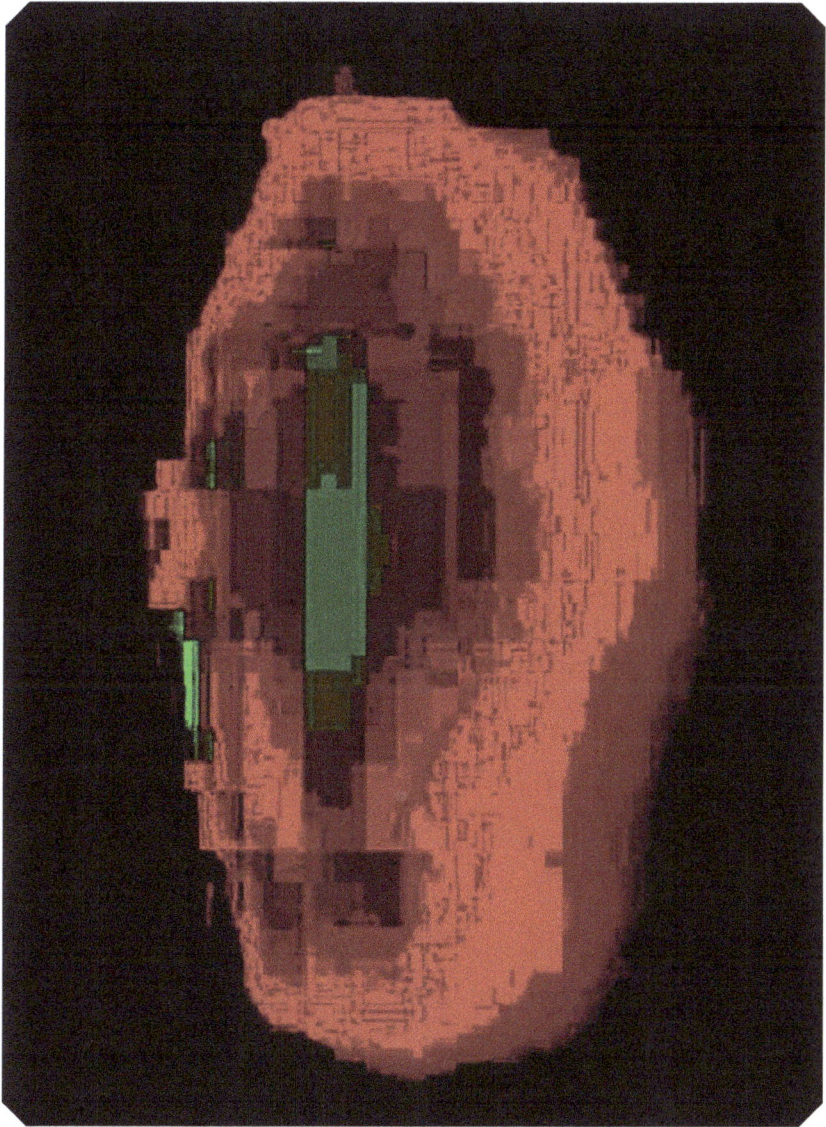

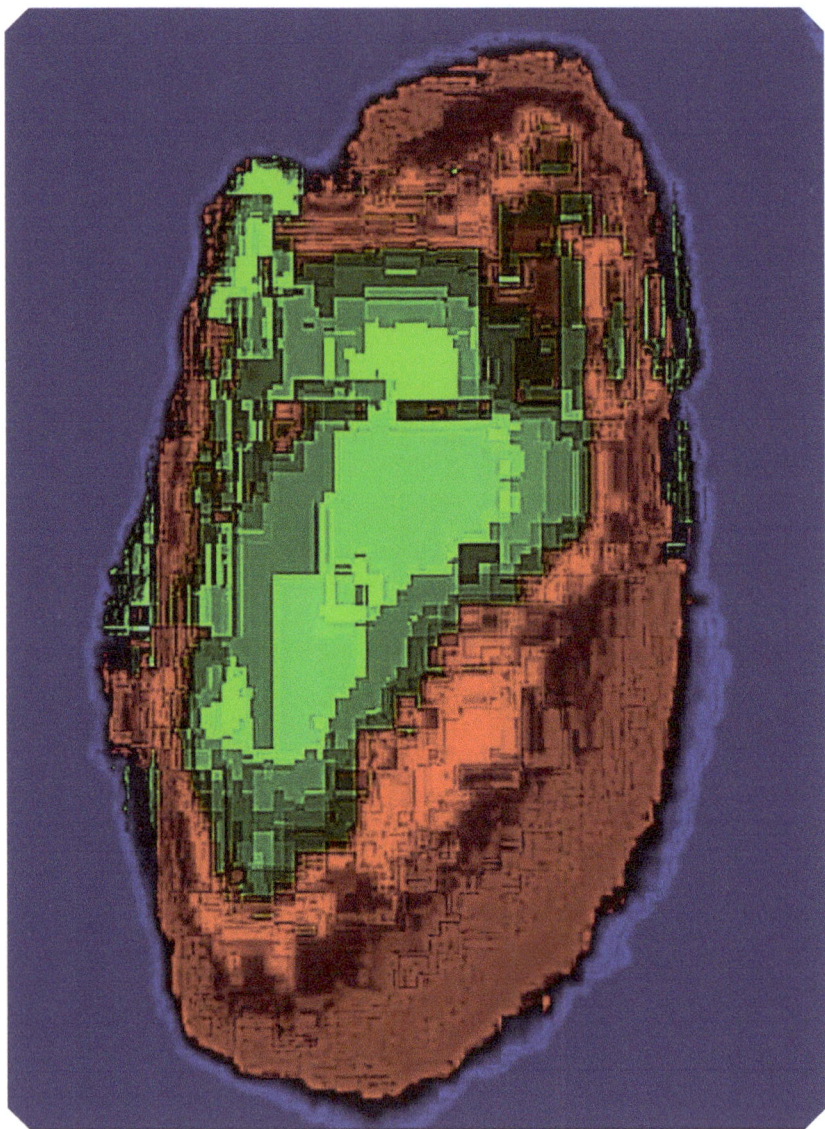

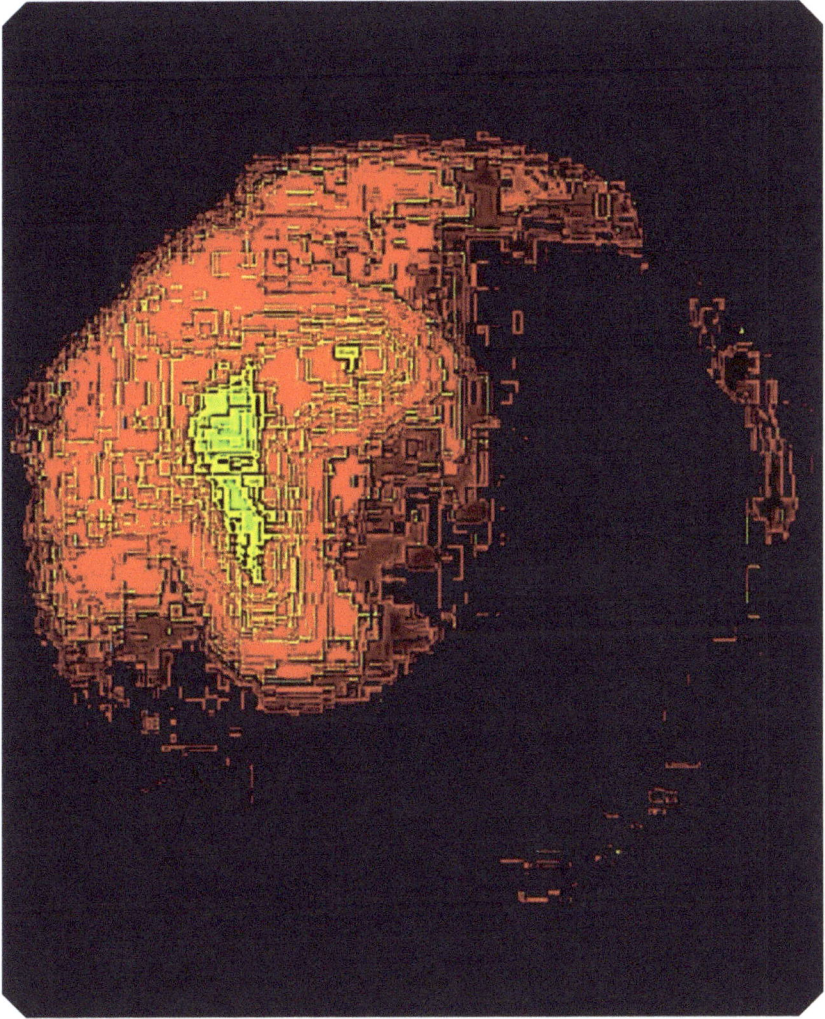

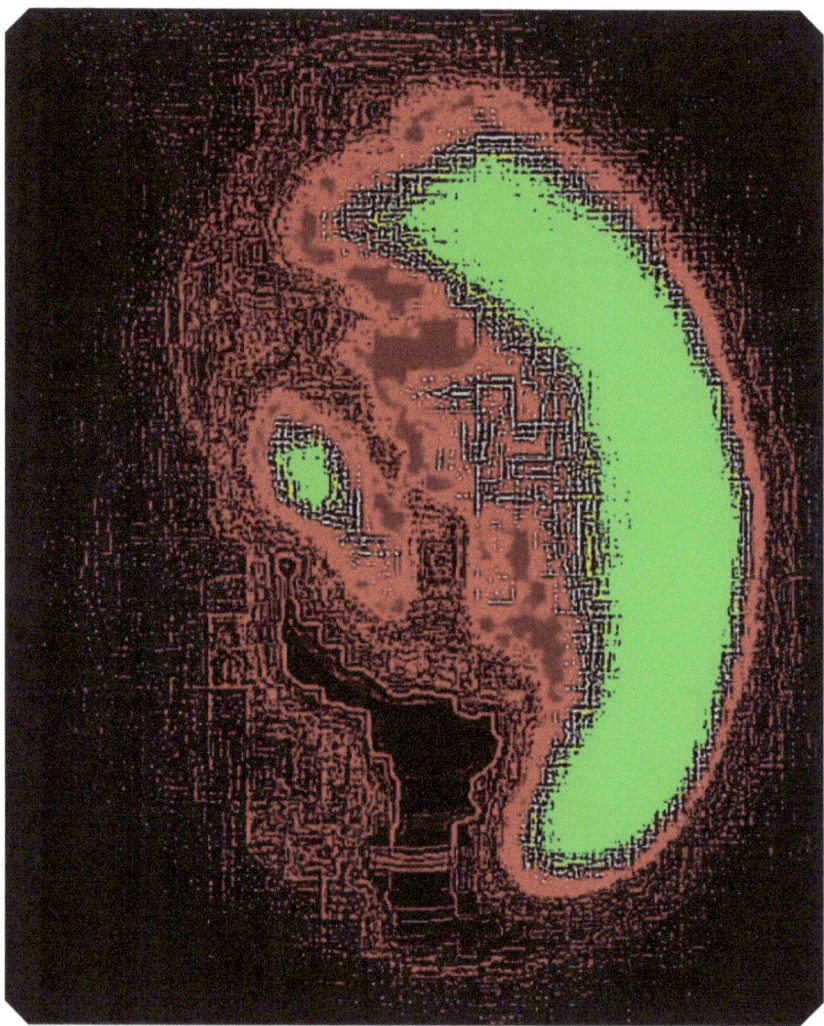

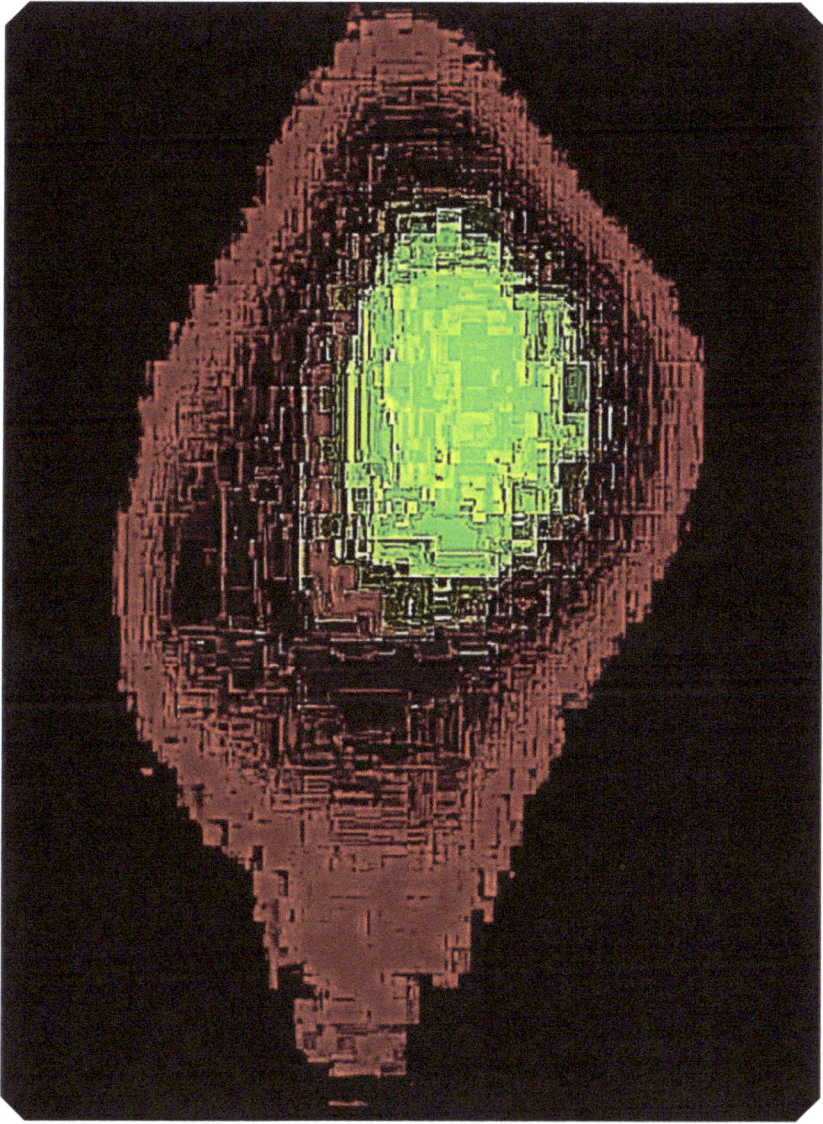

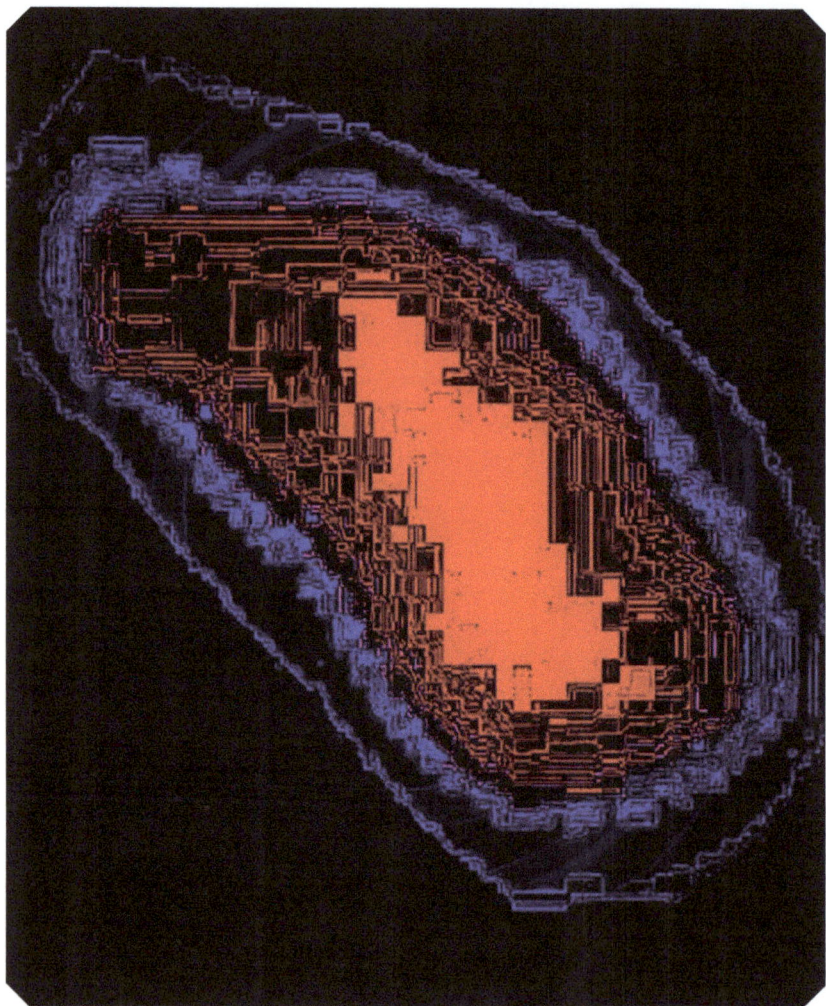

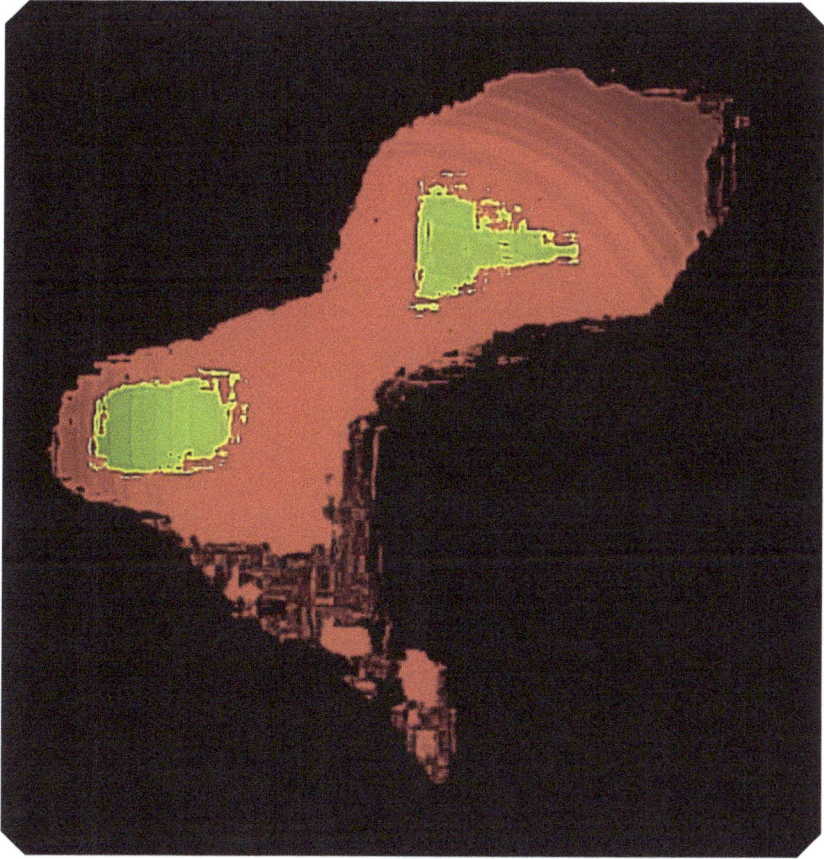

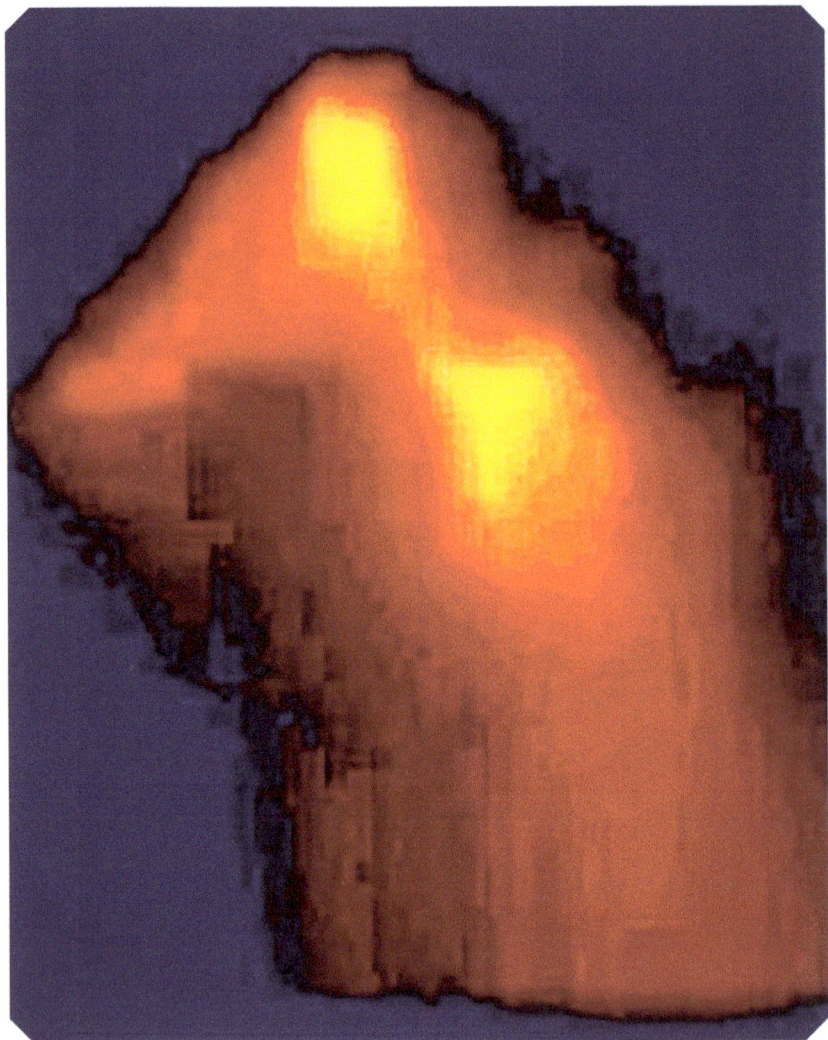

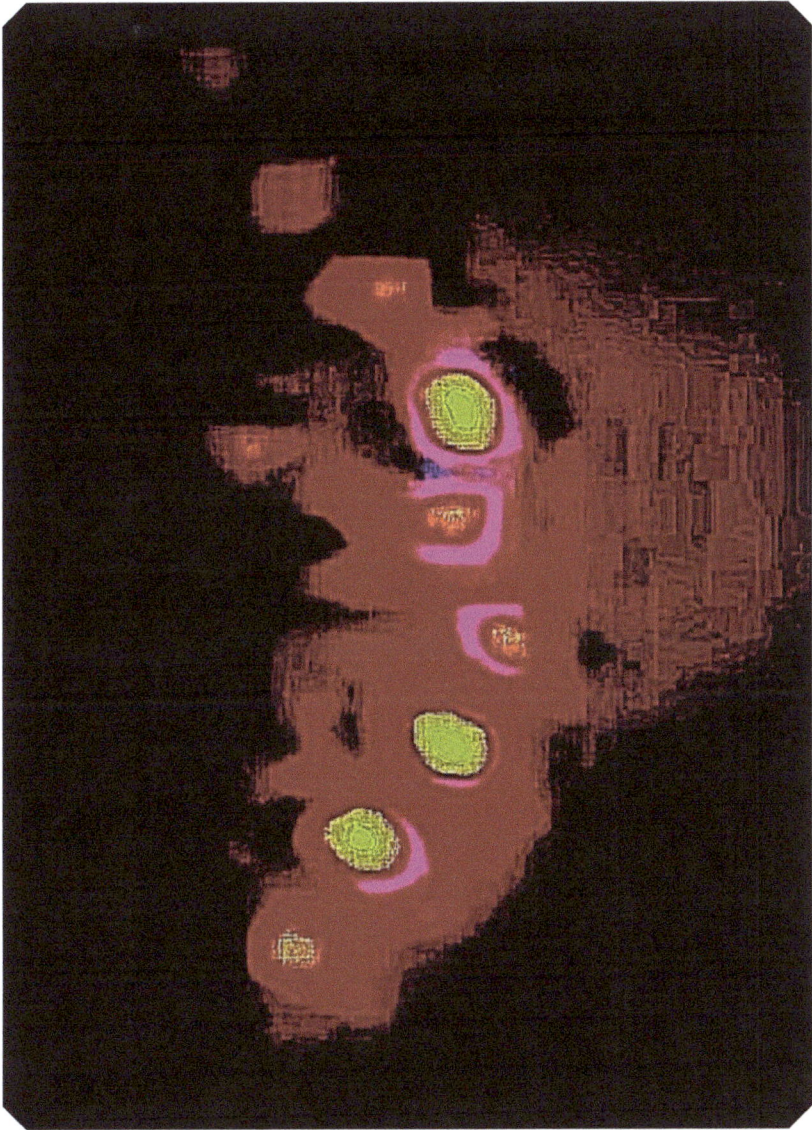

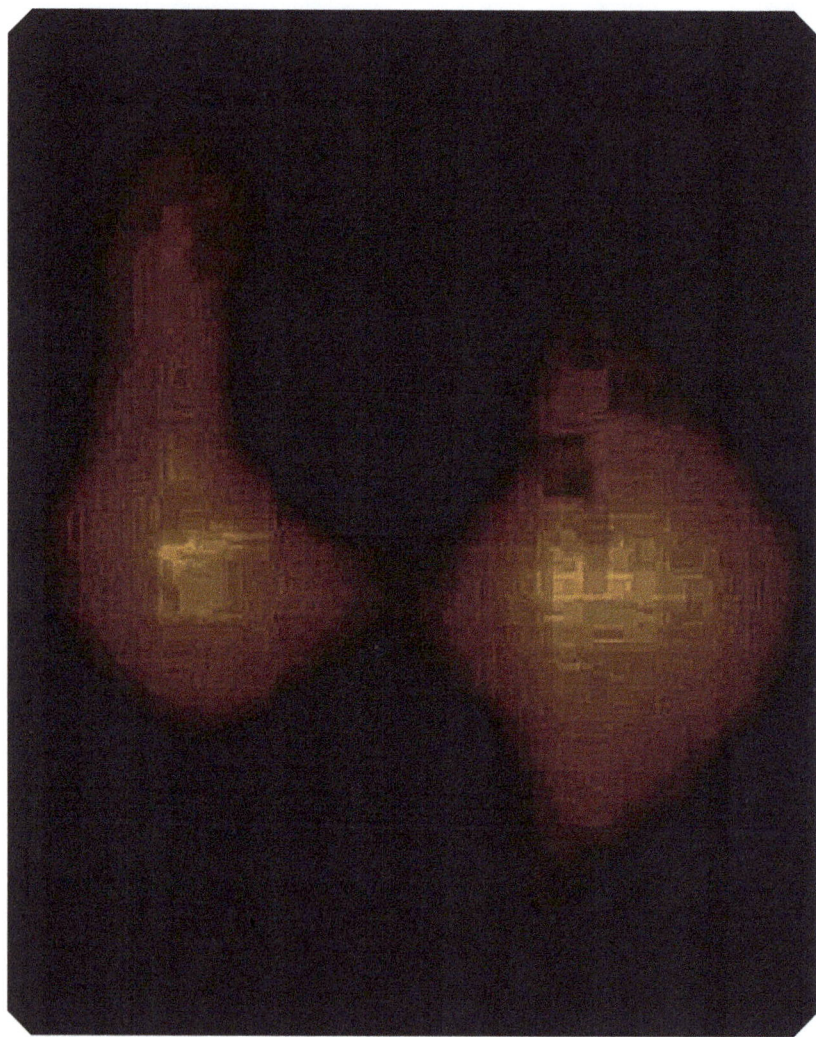

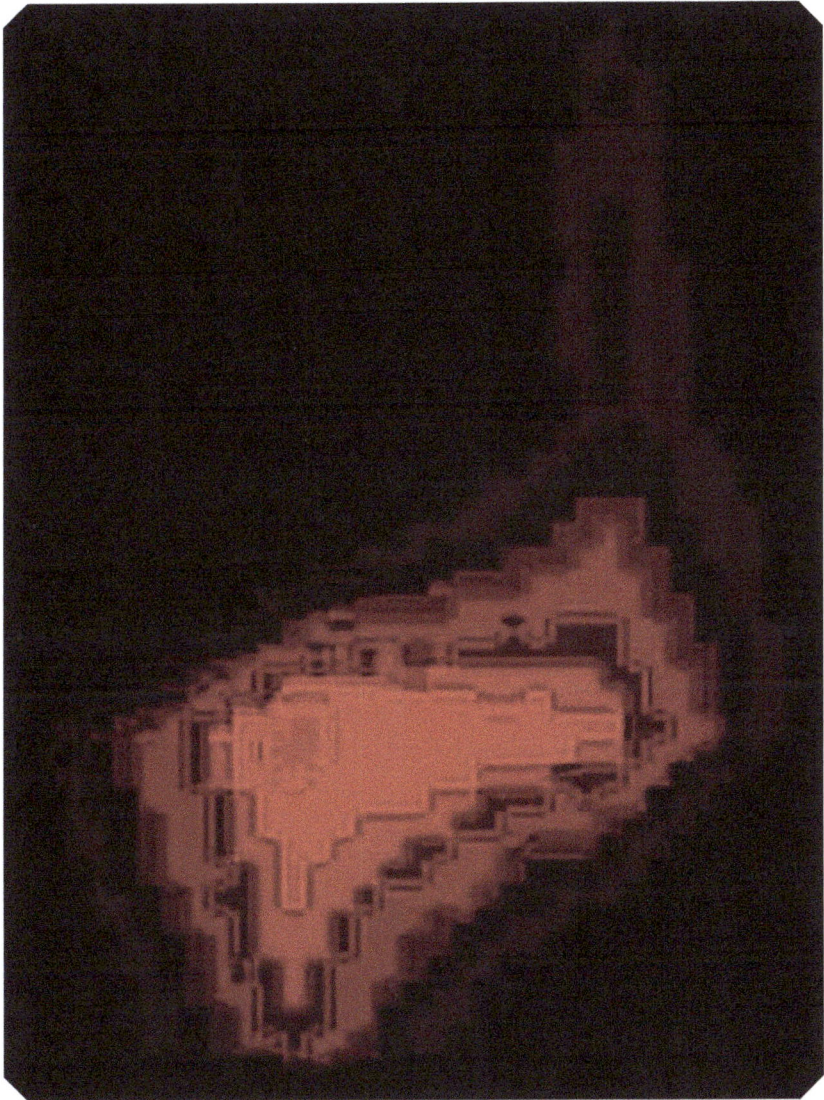

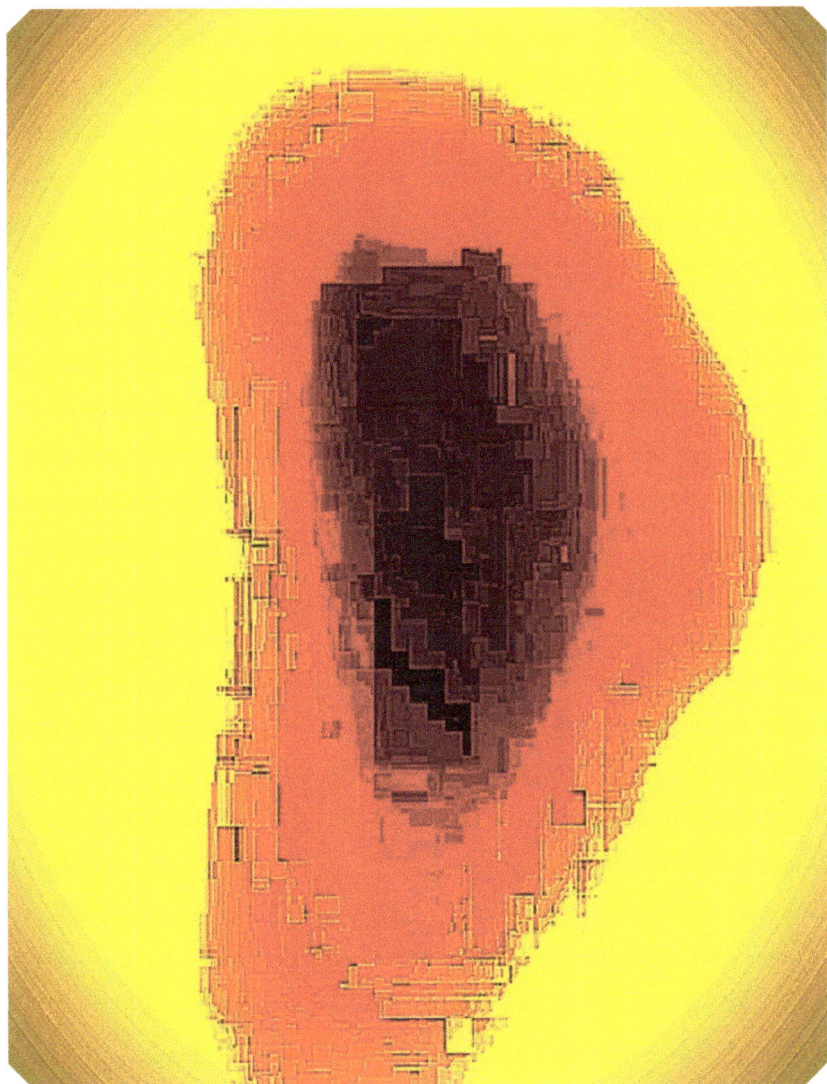

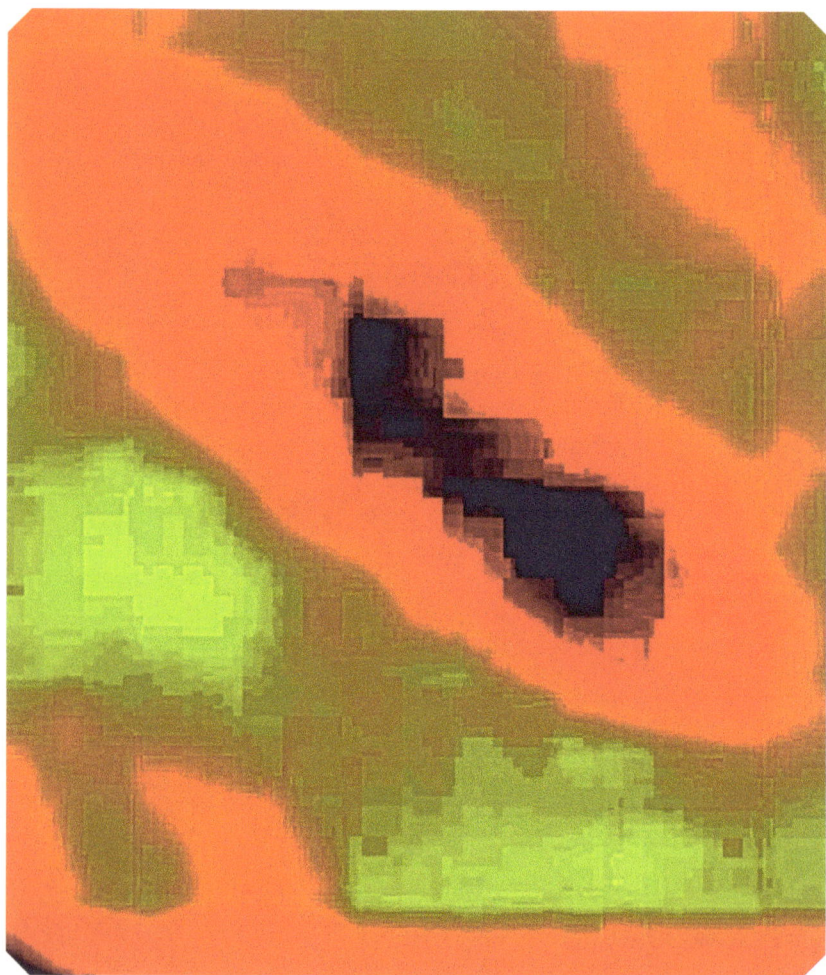

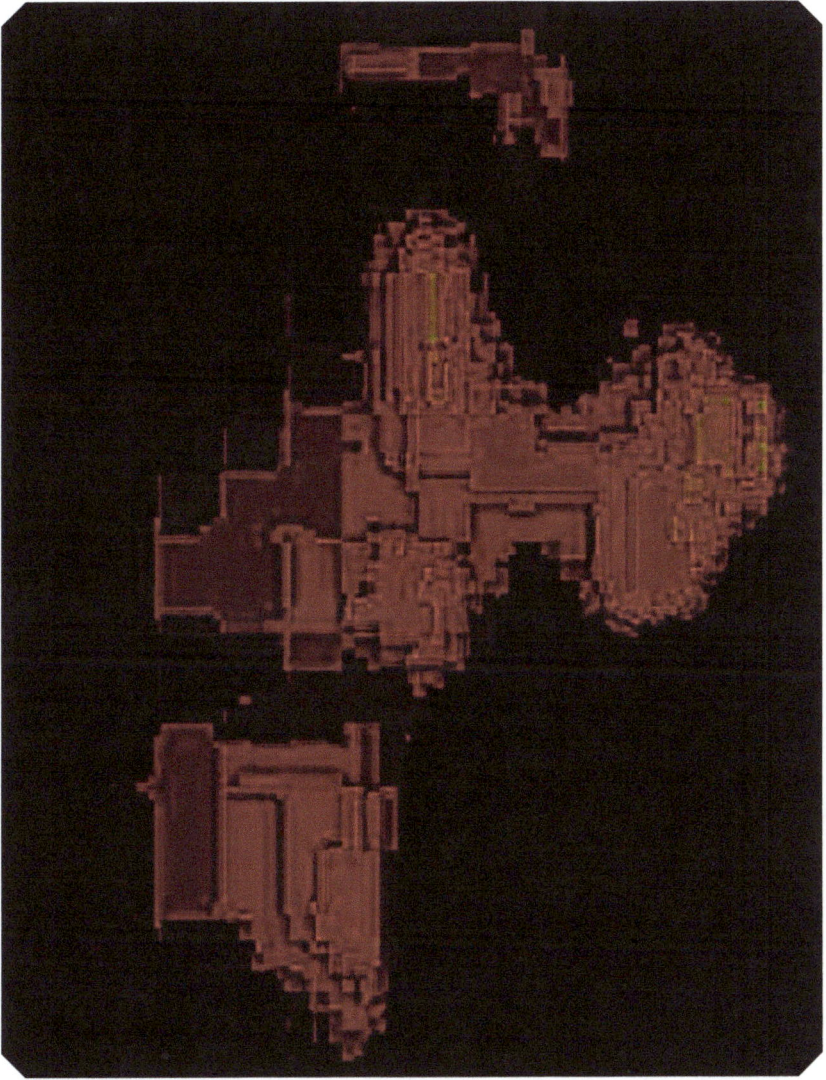

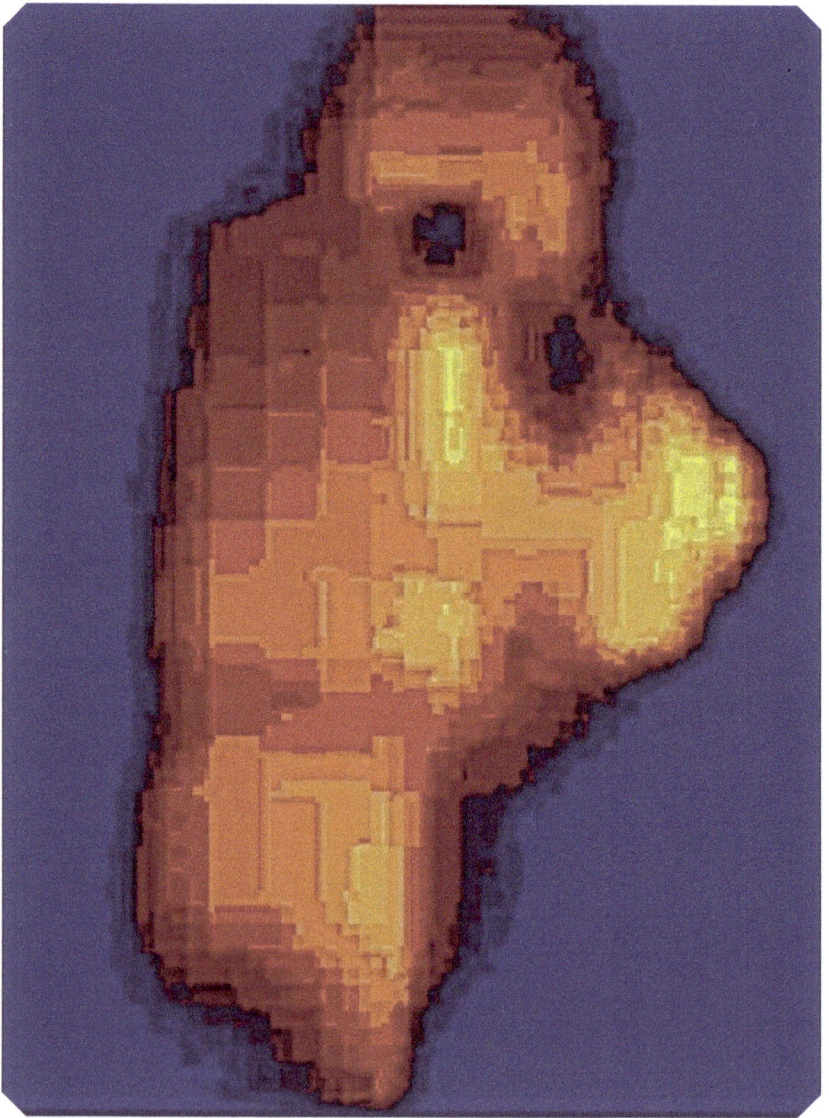

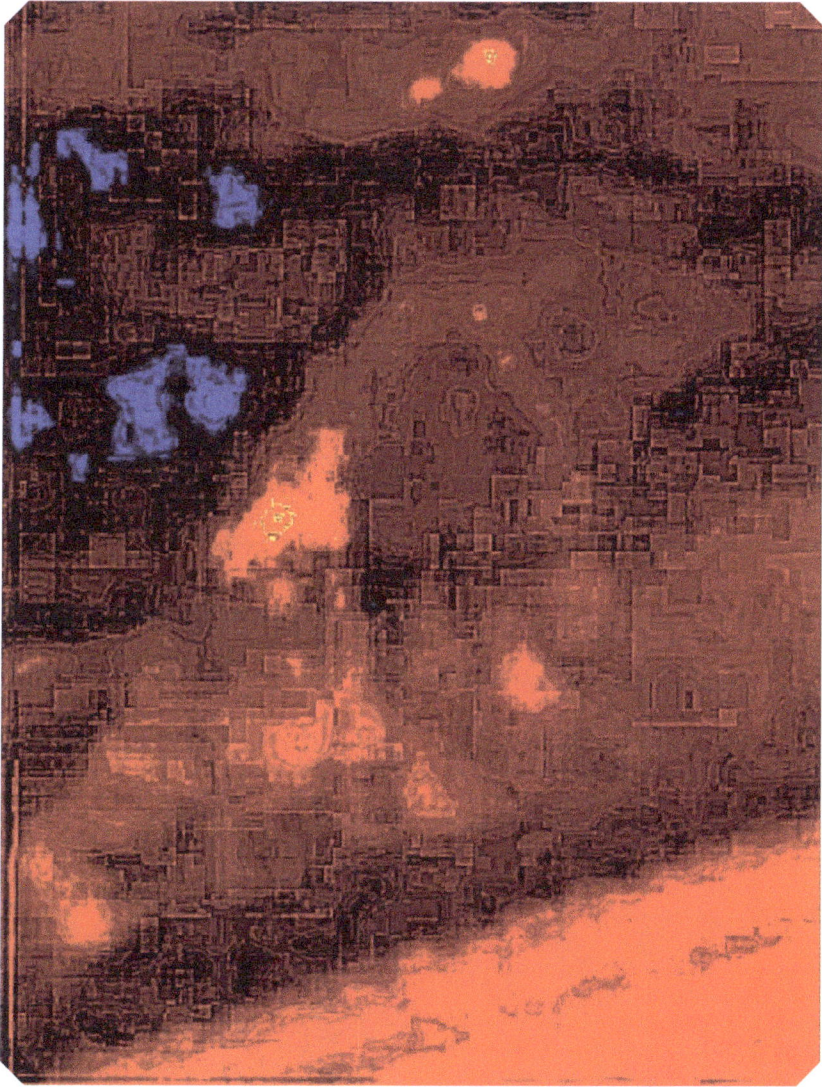

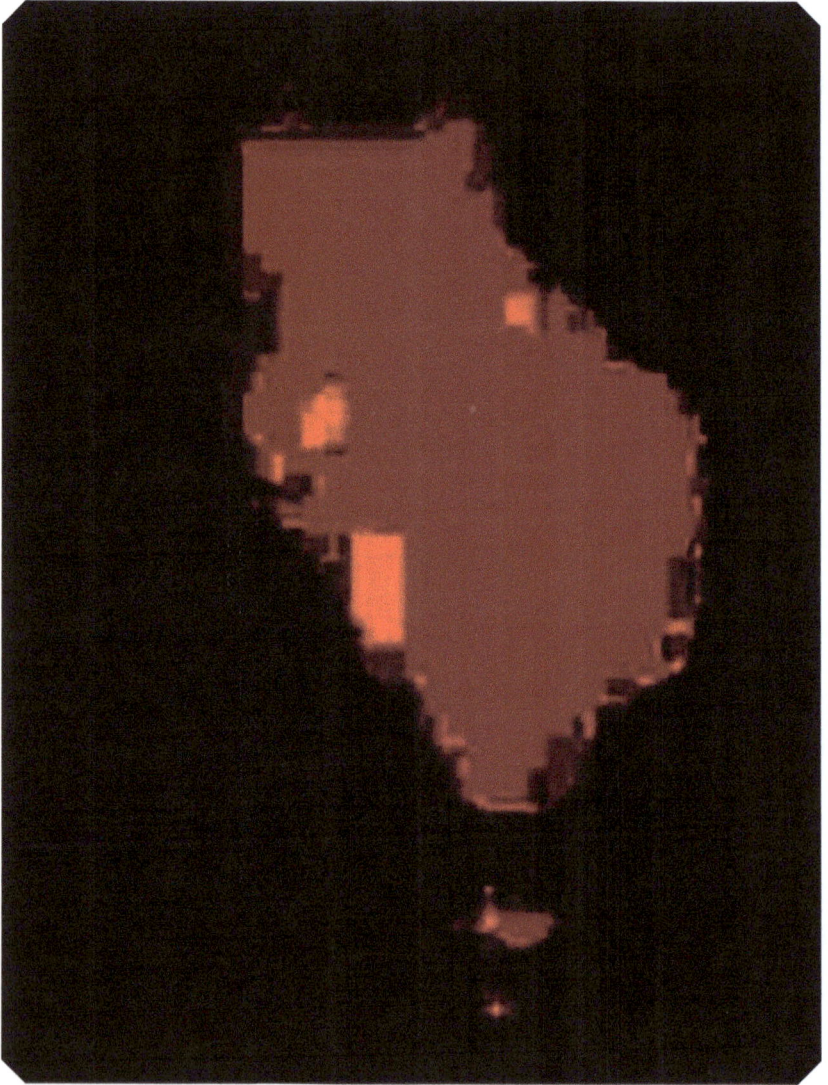

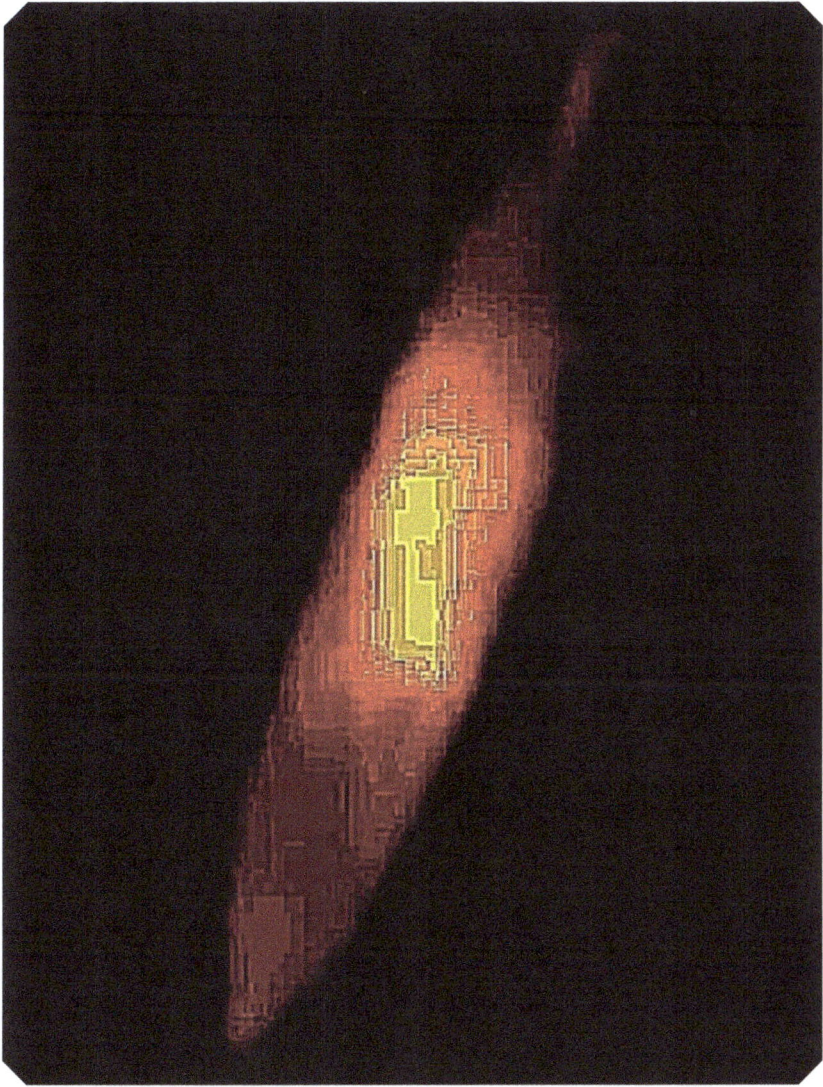

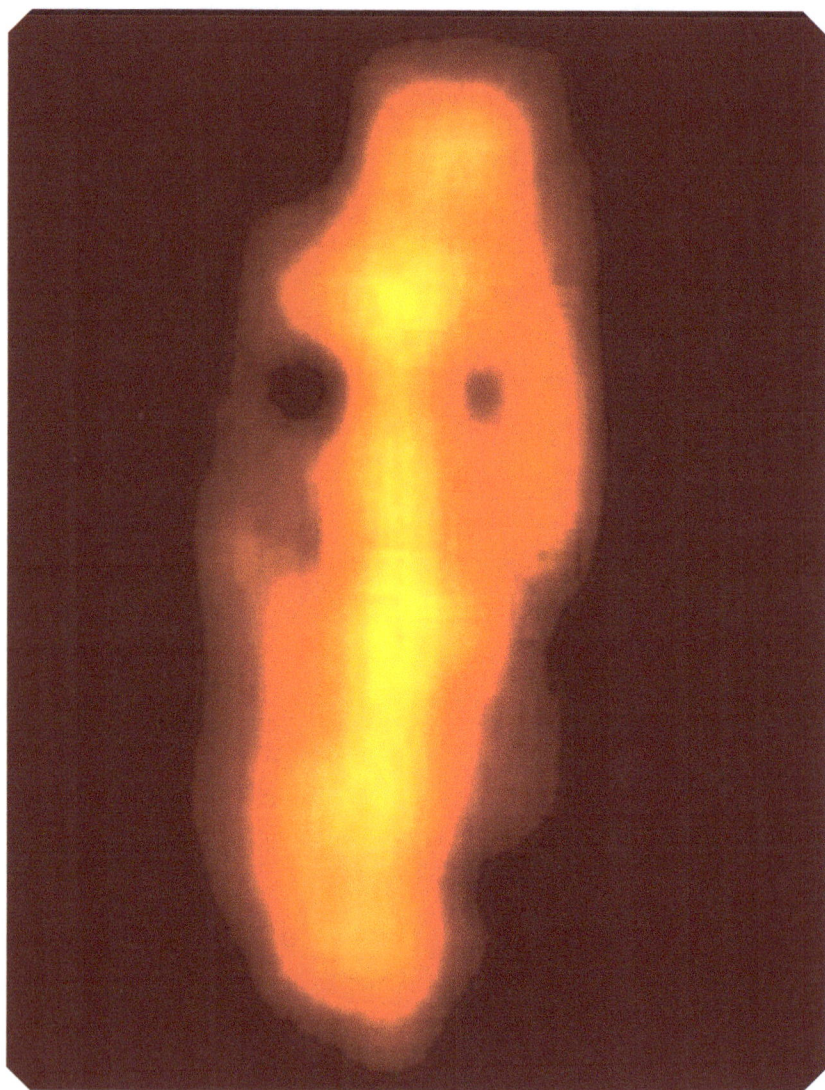

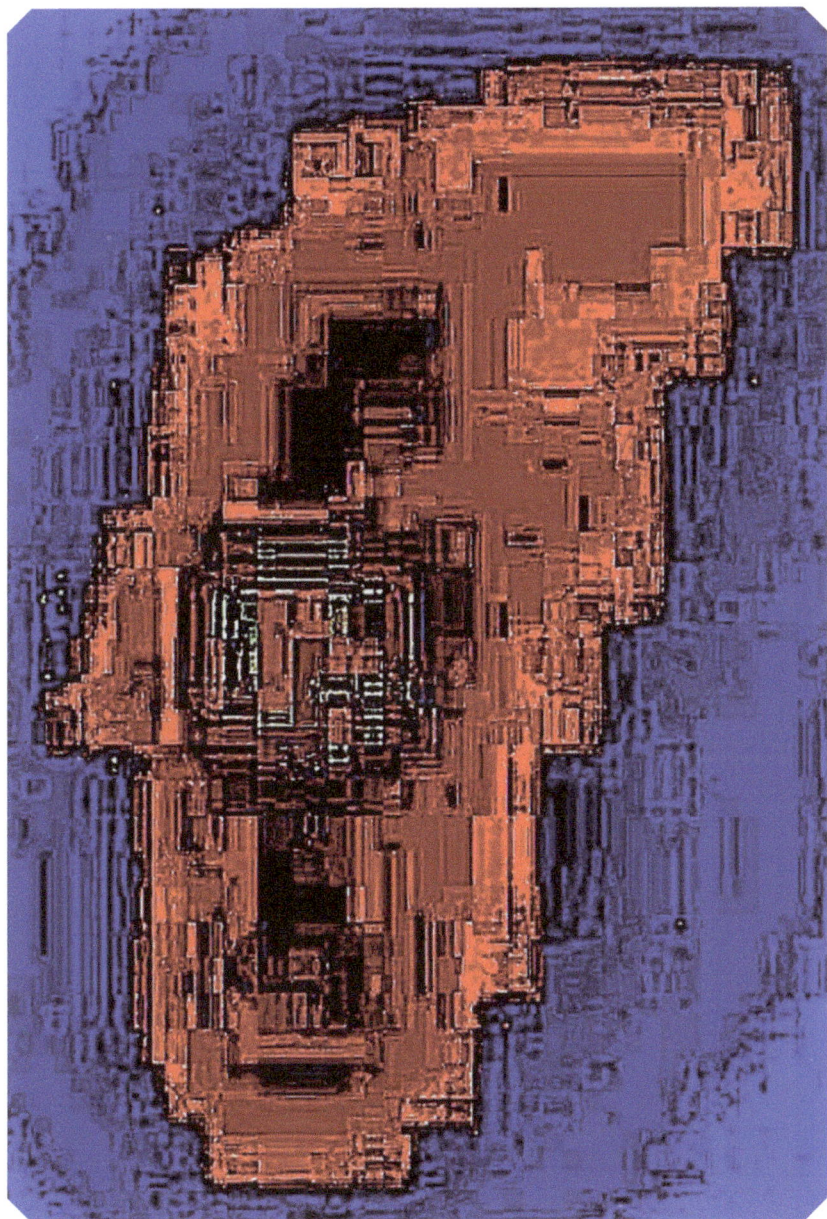

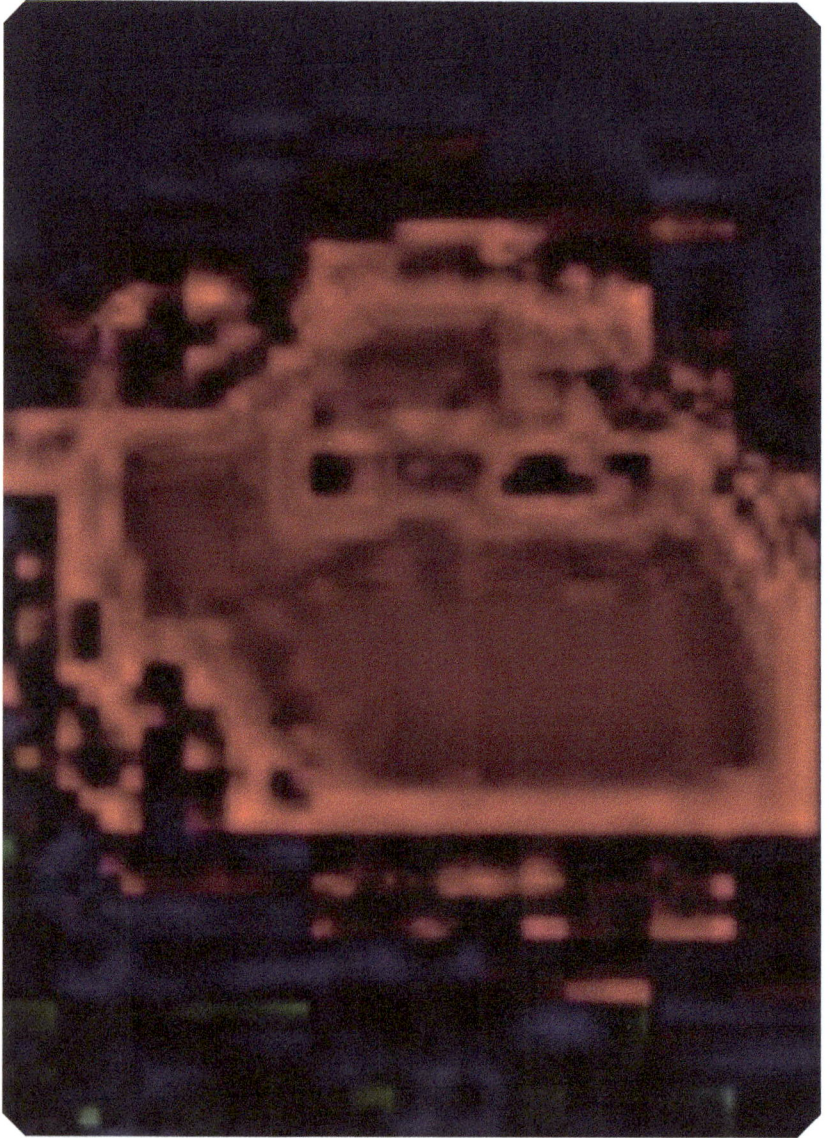

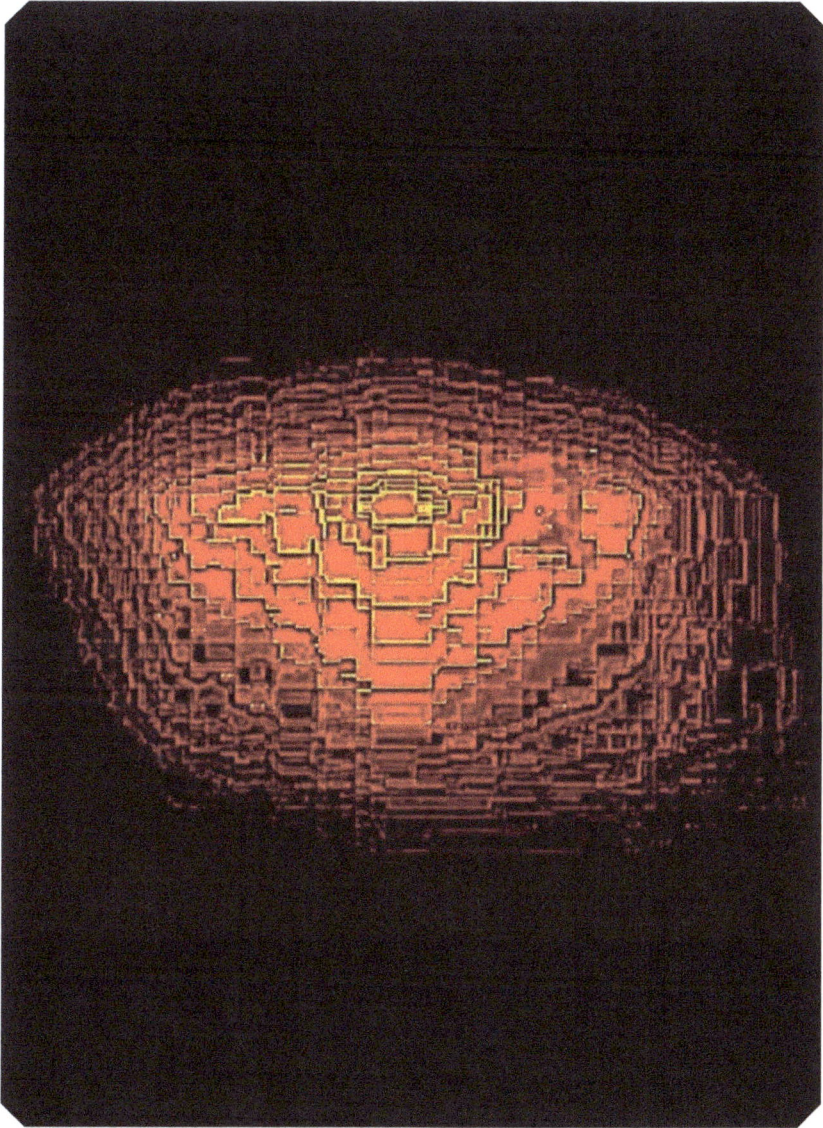

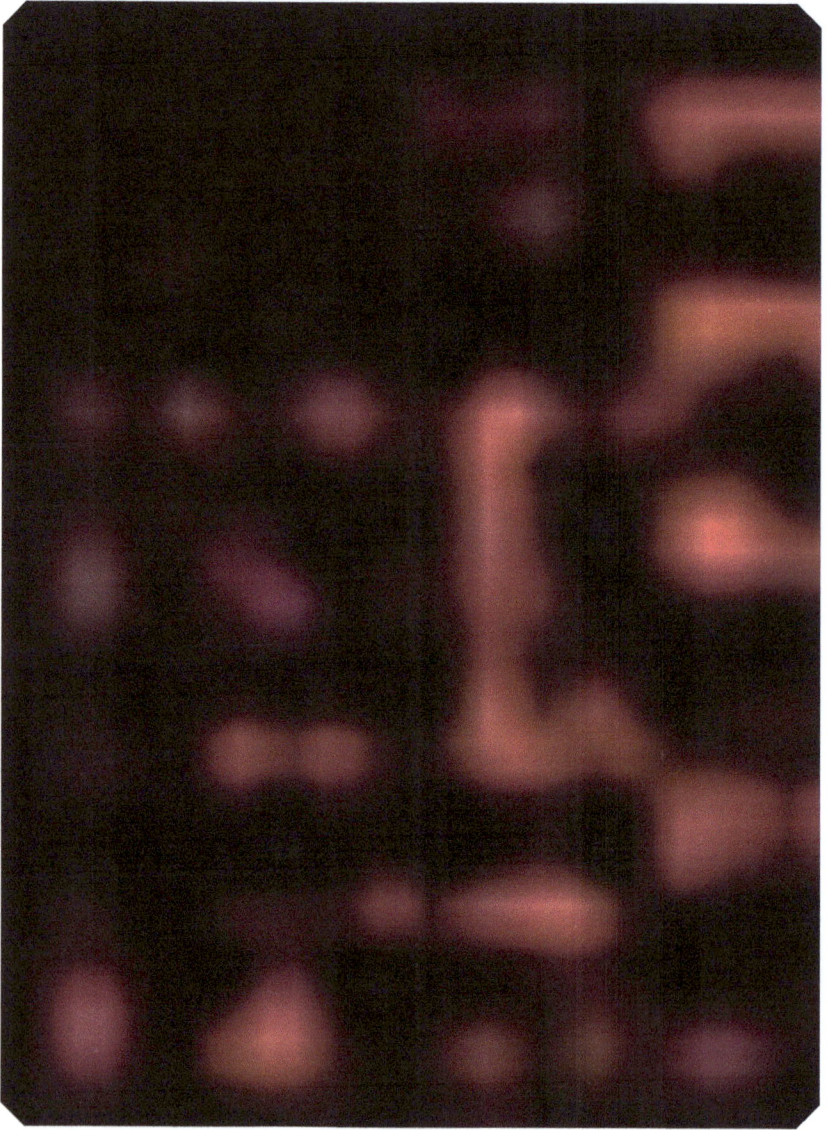

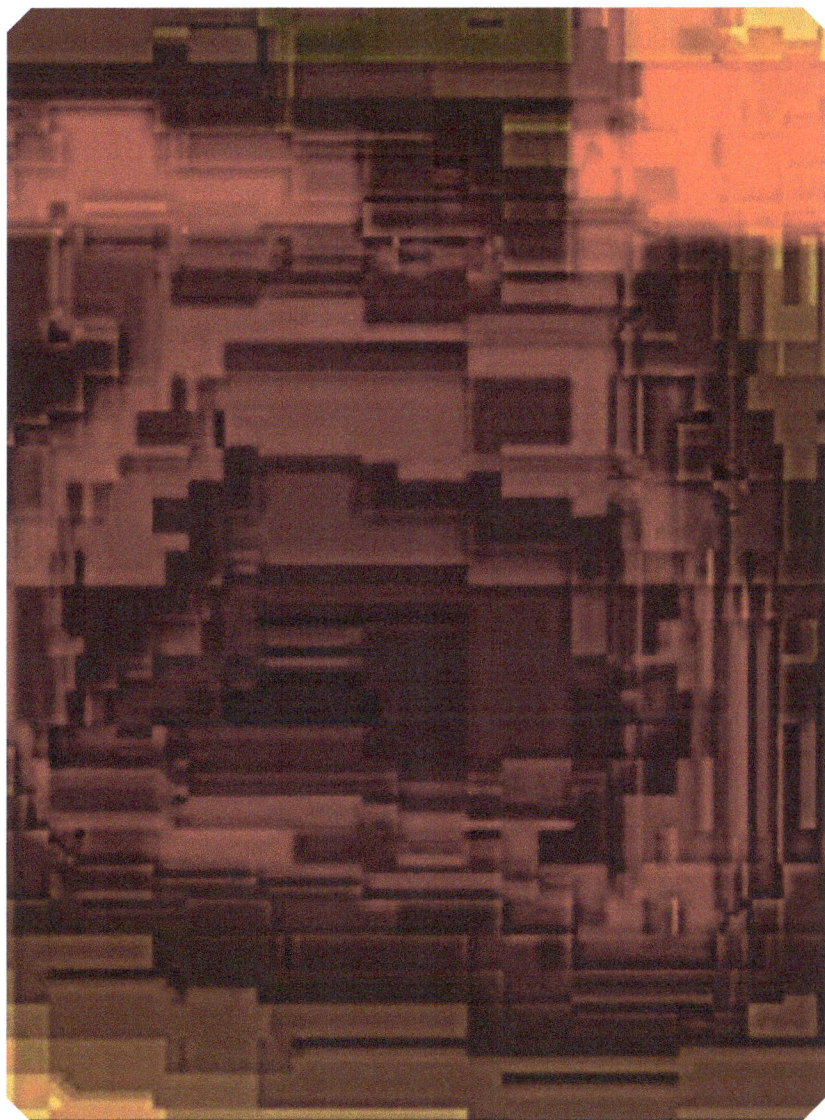

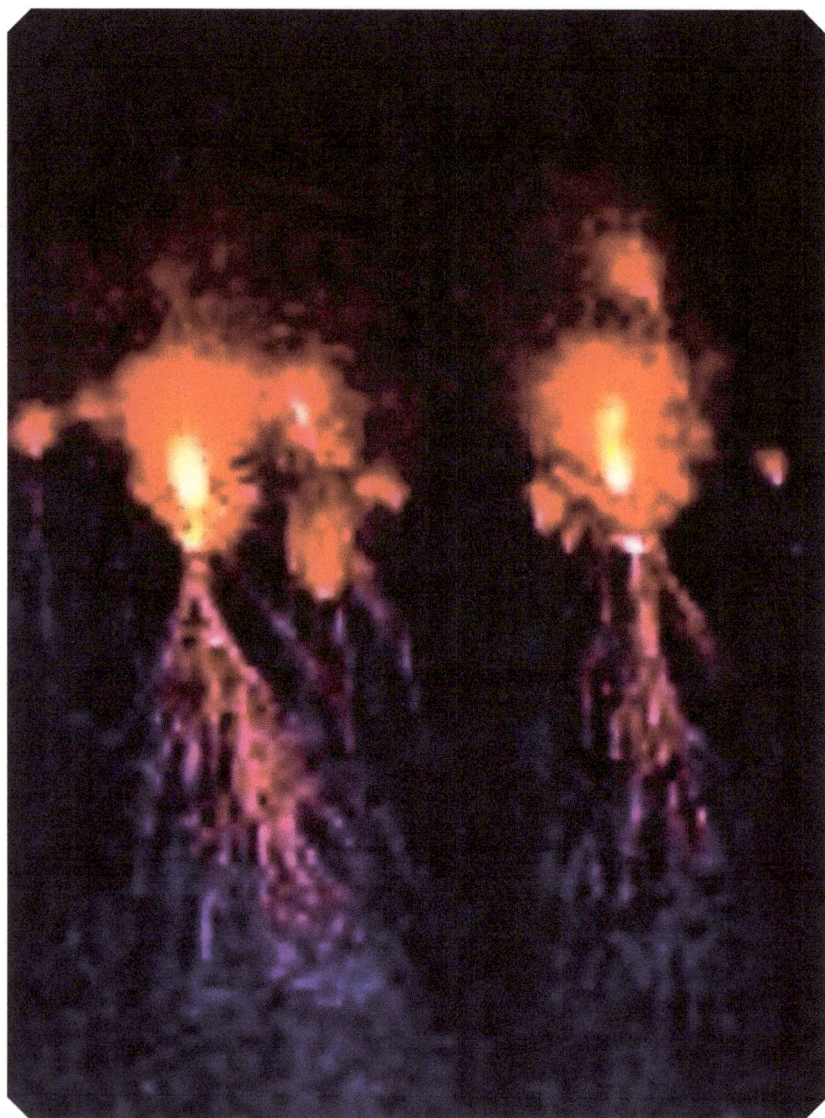

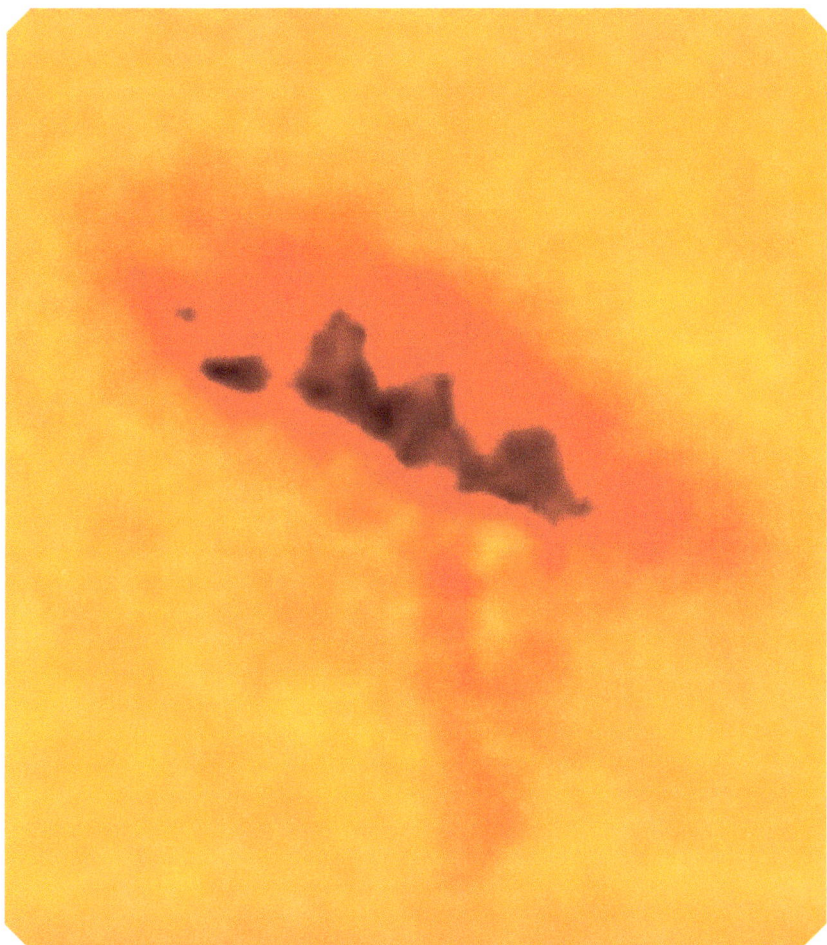

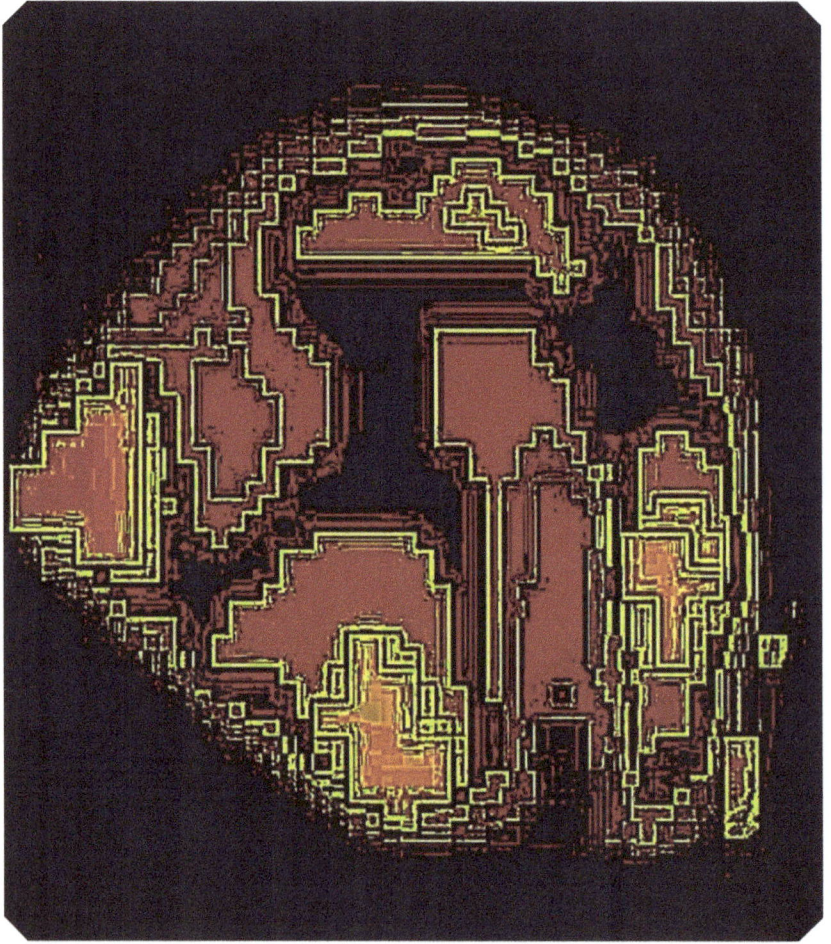

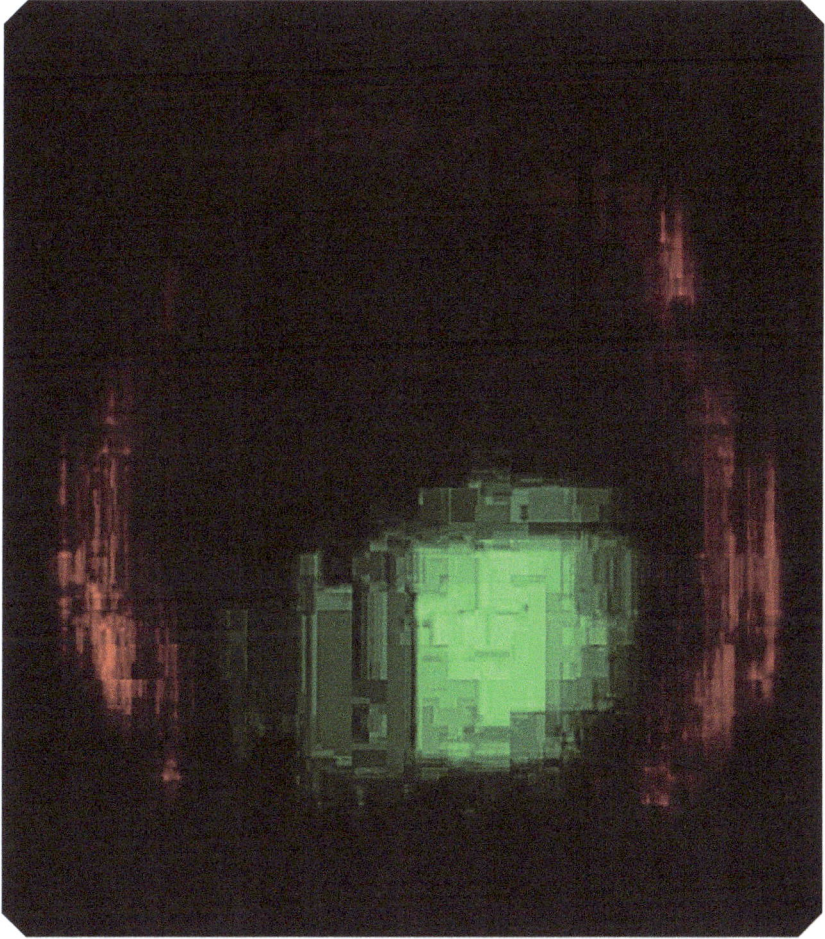

EPILOGUE

I have shown some of the photos in this collection to certain people I know. Many ask me to explain what they are seeing. I simply reply that I have no idea. I am only the person who was able to uncloak them.

The ships are designed and built by civilizations that are far more advanced than human beings, and our looking at them is like showing an automobile to a person living in the Stone Age. They would see the car but would have no idea what they are seeing. Structural engineers may be able to venture a more sophisticated guess as to what the various parts of the starships are, and I welcome their opinions and those of anyone else.

The complete answers to these questions will only come from those who build them and, hopefully, we may be able to ask them directly.

ABOUT THE AUTHOR

Terry Ray does not normally write nonfiction books about UFOs. He was signed with Sunbury Press to write novels, which can be found by searching, "Terry Ray Books." However, an incident in his life caused him to have a life-long interest in UFOs.

When he was ten years old, in the 1950s, Terry was playing football with some the neighborhood boys in an empty lot in the Western Pennsylvania steel mill city where he grew up. The game was interrupted by an enormously loud sound that rumbled the ground under their feet. Suddenly, a shiny, silver, Korean War era jet plane passed low over their heads, chasing another shiny metal aircraft. The aircraft it was chasing was something they had never seen before. It was also made of shiny metal but had a very strange shape and no wings.

The strange aircraft made a sudden turn to the left and headed back, in the direction from which had just come. The jet plane banked sharply to the left and continued its pursuit and they were both quickly out of sight. Terry and his friends were stunned into silence were also very scared. Suddenly, one of the boys broke the silenced by screaming, "That was a spaceship! We have to call the Army!"

Terry's small house was beside the empty lot, so all the boys followed Terry into his house. His mother was in the kitchen, washing clothes, with the drain hose in the kitchen sink. The boys surrounded her, and Terry told her they needed to call the Army. She put her hands on her hips and spoke in a loud voice, "What?"

Terry quickly told her the story of the jet plane and the space-ship. His mother just shook her head. "Well . . . you're not calling them from here so just go back outside and play."

Terry never forgot this experience and was determined to tell the world about it someday. Many years later, after graduating college and serving as an Air Force pilot, Terry became a trial attorney, got married, and had his own family. During a family vacation in 2012 to Ocean City Maryland, Terry was on the balcony of their hotel room, by himself, at about 10:30 at night, looking out at the Atlantic Ocean. The rest of his family was asleep inside the room.

Terry noticed something on the horizon to his left, which was to the north. It was an orange light. This caught his attention because, as a former pilot, he knew that orange was a very unusual color to see in the night sky. The light continued flying south, toward him, along the edge of the beach. It stopped about five miles away. He could see it was a large sphere. Terry estimated that, given the distance, the sphere probably had a diameter about the length of a football field.

It hovered in place for about five minutes and was completely silent, then began moving slowly east, out to sea. It stopped at a distance of about ten miles from shore, and, again, remained there for some minutes before beginning a slow climb to about 10,000 feet in altitude. After another pause, the large orange orb began moving south, toward a large thunderstorm that was out to sea, and flew directly into it and disappeared.

A few minutes after that, Terry noticed another orange light on the northern horizon that proceeded to follow the exact same flight plan as the first orb. The second orb was also silent and the same size as the first one. In all, a total of eight orange orbs, appeared on the northern horizon and flew the same, identical pattern. All were silent and of the same, large size.

This experience prompted Terry to fulfill the commitment he made to himself when he was a young boy of ten . . . he was going to tell the world about this. After more than a year of research and writing, Sunbury Press published his book, *The Complete Story of the Worldwide Invasion of the Orange Orbs*. The book has sold well around the world and is still selling well at the time of this writing.

Being primarily a novelist, Terry decided to incorporate his passion for UFOs into his collection of novels and wrote a fictional extreterrestrial tale entitled *GXM731*, which has also been widely read.

www.ingramcontent.com/pod-product-compliance
Lightning Source LLC
Chambersburg PA
CBHW051213090426
42742CB00021B/3435